William L. Masterton
Department of Chemistry
University of Connecticut

Emil J. Slowinski
Department of Chemistry
Macalester College

ELEMENTARY MATHEMATICAL PREPARATION FOR GENERAL CHEMISTRY

W. B. SAUNDERS COMPANY

Philadelphia, London, Toronto 1974

 Saunders Golden Series

W. B. Saunders Company: West Washington Square
Philadelphia, PA 19105

12 Dyott Street
London WCIA 1DB

833 Oxford Street
Toronto, Ontario M8Z 5T9, Canada

Elementary Mathematical Preparation
for General Chemistry

ISBN 0-7216-6165-3

Last digit is the print number: 9   8   7   6   5   4   3   2   1

PREFACE

The difficulties that students in general chemistry have with elementary mathematics fall into two categories. Many of our students appear to be victims of the "new math." Although quite capable of quantitative reasoning, they are woefully deficient in the basic techniques of mathematics. Frequently we find that a student who can express the principles of chemical equilibrium in terms of an algebraic equation is unable to solve that equation for a numerical answer. Again, a student who has no difficulty explaining what a logarithm is may be totally unfamiliar with the use of a table of logarithms.

Some of our students show a more basic weakness: they have not developed the ability to analyze problems in a logical manner. For these students, translating word problems into mathematical expressions is a major stumbling block. Too often they resort to rote methods under the mistaken impression that it is possible to solve problems without understanding what one is doing.

The predecessor of this book, Mathematical Preparation for General Chemistry, was directed primarily to students in the first category and a review of the basic techniques of mathematics remains a major theme of this book. Exponential numbers, logarithms, the use of the slide rule, and significant figures are discussed in Chapters 4 through 7. Algebraic equations and functional relationships are covered in Chapters 8 and 9. New chapters have been added in two areas where many of our students are deficient: percentage relationships (Chapter 3) and simple techniques of graphing (Chapter 10).

A major feature of this book is the addition of material directed to the student who has difficulty "setting up" problems in general chemistry. The first (and longest) chapter considers a systematic approach to problem analysis. Emphasis is placed on the importance of recognizing and understanding the chemical principles involved. This theme is further developed in Chapter 2, where the unit conversion approach is discussed. New material has been added to Chapter 9 to show how chemical principles are translated into algebraic equations, with particular emphasis on problems dealing with chemical equilibrium. We are hopeful that these additions will help those students who have had little training in problem analysis prior to the college level course in general chemistry.

Certain more advanced topics which appear in Mathematical Preparation for General Chemistry have been deleted from this text. These include chapters on geometry, trigonometry, and calculus, which are seldom required for the first year course. We have, however, retained the discussion of error analysis (Chapter 11) as applied to the chemistry laboratory.

We have included a large number of worked examples and problems for which answers are provided. The application of mathematical techniques to chemistry is stressed throughout; nearly all of the problems are phrased in terms of chemical principles. It is not intended that the several chapters be "covered" or "assigned" in the usual sense. Rather, we would expect the student who is having difficulties with stoichiometric calculations to become familiar with the conversion factor approach by reading and working the problems in Chapter 2. Students who are unable to work "pH problems" can be referred to the treatment of logarithms in Chapter 5. We hope that in this manner the instructor can spend more time on chemistry and less on the elementary mathematics that students should have mastered (but often have not) before they enter college.

WILLIAM L. MASTERTON
EMIL J. SLOWINSKI

CONTENTS

PROBLEM ANALYSIS

Most of this book is devoted to the mathematical techniques used to work out answers to problems in general chemistry. We will consider such operations as expressing numbers in exponential notation, taking logarithms of numbers, and solving algebraic equations. As important as these techniques are, you will find them of little or no value unless you develop the ability to analyze and set up problems. In this introductory chapter, we will be interested in developing a logical approach to problem analysis. We will suggest some general principles to help you translate the written statement of a problem in chemistry into a simple mathematical expression which you can solve.

To illustrate what this chapter is all about, let us suppose that you are confronted with the following problem. (It, or one of its close relatives, shows up quite frequently on examinations in general chemistry.)

A solution of 1.25 grams of a certain nonelectrolyte in 26.8 grams of water freezes at −1.40°C. *What is the molecular weight of the nonelectrolyte?*

If you analyze this problem correctly, you may come up with the following equation, where M is the molecular weight:

$$1.40 = \frac{1.86 \times 1.25}{M \times 0.0268} \tag{1.1}$$

Probably you can solve this equation for M. Applying a simple principle of algebra (Chapter 8), using a slide rule (Chapter 6) to carry out the arithmetic, and expressing the answer to the correct number of significant figures (Chapter 7), you should arrive at a molecular weight of about 62.0. Most of your fellow students can do this, too; *provided they can come up with Equation 1.1*, they can obtain a numerical answer. However, experience suggests that a large number of students in general chemistry will have trouble going from the statement of the problem to Equation 1.1. For them, "setting up" the problem is the stumbling block. Hopefully, this is not true in your case. If you can see immediately how the equation follows from the statement, the chances are you can afford to skip this chapter. If not, you may wish to refer to Example 1.3, where the thought process involved in analyzing this problem is discussed at some length. Even better, read the whole chapter; you may find that it will sharpen your skills in problem analysis.

1.1 GENERAL APPROACH

Hundreds of years ago, alchemists devoted their time and energy to a fruitless search for the "philosopher's stone" with which they could convert base metals to gold. In a somewhat similar way, generations of students have searched for a "magic formula" that will enable them to solve any and all problems in general chemistry. Their efforts have been even less rewarding than those of the alchemists. Although we would like to be able to furnish you with a universal method of solving general chemistry problems, we must admit that even after twenty years of teaching and studying chemistry, we have yet to discover any magic relations. The simple fact is that if you hope to get an A in this course, or even a B for that matter, (probably even a C), you simply must understand what you are doing when you solve problems.

Although there are no magic formulas for success in chemistry, we can suggest some general rules which should prove helpful in analyzing problems. In this section, we will consider five such rules. Quite possibly, your instructor can suggest others, equally important, that we have overlooked. Indeed, you may be able to add to this list yourself as you gain experience and insight.

1. **Read the problem carefully; make sure you know what is given and what is asked for.**

Anyone who has graded examinations in general chemistry can cite cases where a student's answer does not correspond to the question asked. Presumably, he failed to read the question carefully. A typical example, of a nonmathematical nature, is shown in Question A, Figure 1.1. What grade would you give for this answer? Chemically, it is correct; had the student been asked to prepare NH_4Cl from NH_3, he would clearly have deserved full credit. Unfortunately, he was asked the reverse question: how to prepare NH_3 from NH_4Cl.

Question A: How would you prepare ammonia, NH_3, from ammonium chloride, NH_4Cl?

Student Answer:

Saturate a solution of hydrochloric acid with ammonia and evaporate to obtain crystals of NH_4Cl.

Question B: A sample of a certain gas occupies a volume of 210 cc at 300°K. At what temperature, in °C, will it have a volume of 250 cc?

Student Answer: $T_2 = T_1 \times \dfrac{V_2}{V_1} = 300°K \times \dfrac{250\,cc}{210\,cc} = 357°K$

FIGURE 1.1 Make sure you read the question!

Another example, this time a little more subtle, appears in Question B of Figure 1.1. Here again, everything that the student has written down is correct. He has related temperature and volume correctly, his arithmetic cannot be faulted, and indeed the final temperature is $357°K$. However, he seems not to have noticed that the temperature was required in $°C$. He should have gone one step further and applied the relation:

$$°K = °C + 273$$

to obtain a final answer of $84°C$.

An instructor looking over the answers to these questions might become a trifle suspicious. Could it be that the student, unable to answer the question in the form it was asked, deliberately modified it to a form he could deal with? This type of gamesmanship is not unheard of in general chemistry. How successful it is depends largely upon the instructor's sense of humor and the extent to which he appreciates originality in his students. It is probably safest not to try to outmaneuver him; *answer the question that was asked.*

2. Make sure you understand all the terms used in the statement of the problem.

Occasionally, you may find yourself "hung up" on a problem because you do not understand a word or phrase used in its statement. Sometimes it is clear that the answer hinges on the meaning of a particular word. This is the case with the multiple-choice questions of Figure 1.2, which really involve little more than definitions. Clearly, you can't answer these questions correctly unless you know what the words *density* and *mole* mean. More exactly, according to the laws of probability, you have one chance in five of guessing the correct answer. Since 20% is seldom an acceptable grade, you should make sure that you are thoroughly familiar with terms such as these; they are discussed in your text.

It is not enough that you be able to rattle off definitions of scientific terms. On homework assignments and examinations, you are seldom asked for definitions as such. Instead, you will be expected to answer questions that assume a precise, working knowledge of one or more scientific terms. There is a simple way to check the true extent of your scientific vocabulary. Try to explain orally what is meant by such terms as *mole, atomic weight, heat of formation*, citing examples. When you can do this with no false starts or "hemming and hawing," you are ready for the most important step in problem analysis.

Question 1: A piece of metal weighing 6.40 grams occupies a volume of 3.94 cc. The density of the metal may be found by:
   a) Multiplying 6.40 g by 3.94 cc.
   b) Dividing 3.94 cc by 6.40 g.
   c) Dividing 6.40 g by 3.94 cc.
   d) Adding 6.40 g and 3.94 cc.
   e) Subtracting 3.94 cc from 6.40 g.

Question 2: The mass of two moles of H_2O is:
   a) 18.0 g   b) 9.0 g   c) 36.0 g   d) 12.0×10^{23} g   e) 3.0×10^{-23} g

   (The correct answer in each case is *c*)

FIGURE 1.2 Importance of knowing the meaning of terms.

3. **Decide what chemical principle is involved and how you will use it to set up the problem.**

Every problem in general chemistry involves applying one or more of the principles you have learned. These principles are stated and described in your textbook; in addition, your instructor will ordinarily discuss and illustrate them in lecture. It is essential that you become completely familiar with these principles before you attempt to work problems based upon them. Otherwise, you will be unable to recall the proper principle and apply it to the case in hand. Indeed, this is the critical step in the analysis of any problem, mathematical or otherwise.

Sometimes, as in the following example, the problem can be solved by applying a single principle. In this particular case, the problem is virtually solved once you recognize that principle.

Example 1.1   Given the balanced equation:

$$2NO(g) + O_2(g) \rightarrow 2NO_2(g)$$

calculate the number of moles of NO_2 that can be produced from 1.43 moles of O_2.

Solution   To solve this problem, we require a relationship between moles of O_2 and moles of NO_2. That relationship becomes obvious when you recall the general principle that *the mole ratios of reactants and products are given by the coefficients of the balanced equation.* Since the coefficients here are 1 for O_2 and 2 for NO_2, it follows immediately that two moles of NO_2 are formed for every mole of O_2 that reacts. In this case, the number of moles of NO_2 produced must be:

$$2 \times 1.43 = 2.86$$

Many times, we must admit, a problem does not "unravel" as simply as Example 1.1 after you have spotted the appropriate chemical principle. In order to apply the principle correctly, you must understand clearly what is meant by each term involved. The following example sometimes gives students trouble because they have only a hazy idea of what is meant by a mole.

Example 1.2   Given the balanced equation:

$$2NO(g) + O_2(g) \rightarrow 2NO(g)$$

calculate the number of grams of NO_2 that can be produced from 2.46 moles of O_2.

Solution   Note the subtle but important distinction between this problem and Example 1.1. You are asked to calculate the number of *grams* of NO_2 rather than the number of *moles*. The principle involved is the same; indeed, the first step in the analysis repeats Example 1.1. The coefficients of the balanced equation tell us that two moles of NO_2 are formed from one mole of O_2. It follows that we obtain:

$$2 \times 2.46 = 4.92 \text{ moles of } NO_2$$

To complete this problem, you must convert moles of NO_2 to grams. This conversion requires that you clearly understand what is meant by the term "mole." Specifically, you must recognize that the number of grams of NO_2 can be obtained by multiplying the number of moles by the molecular weight, M.

$$\text{no. of grams} = (\text{no. of moles}) M$$

In this case, where the molecular weight of NO_2 is 46.0 [i.e., A.W. of N + 2(A.W. of O) = 14.0 + 2(16.0) = 46.0], the number of grams of NO_2 produced must be:

$$4.92 \text{ moles} \times 46.0 \text{ g/mole} = 226 \text{ g}$$

Notice that to solve this problem, it is not sufficient that you be able to *define* the mole. You must have progressed beyond this point to become thoroughly familiar with the mole concept. The two-step path* that we have described here is unlikely to occur to a student who has only a vague idea of the relationship between moles and grams of a substance.

Many of the relationships discussed in general chemistry are best expressed by means of simple algebraic equations. We cannot emphasize too strongly that *you must understand clearly the meaning of each symbol used in that equation* (Example 1.3). Otherwise, your analysis of the problem will degenerate into mere number-juggling.

Example 1.3 A solution of 1.25 grams of a certain nonelectrolyte in 26.8 grams of water freezes at $-1.40°C$. What is the molecular weight of the nonelectrolyte?

Solution The principle here should be familiar to any student who has read that section of his textbook dealing with the colligative properties of solutions: *the freezing point lowering is directly proportional to the molality of the solute.* For water solutions, where the proportionality constant is $1.86°C$, this relationship is expressed by the equation:

$$\Delta T_f = (1.86°C)m$$

where ΔT_f = freezing point lowering, m = molality.

---

*A convenient way of setting up problems of this type is discussed in Chapter 2.

Most students get to this point in their analysis of the problem. However, they are often stymied as to where to go from here. In order to apply this equation to the problem at hand, you must know precisely what is meant by the terms ΔT_f and, in particular, m. Let us consider each of these terms in succession.

1. The *freezing point lowering* is the difference between the freezing point of the solvent and that of the solution. In this case, since pure water freezes at $0°C$, we have:

$$\Delta T_f = 0 - (-1.40°C) = 1.40°C$$

2. The *molality* is defined as the number of moles of solute per kilogram of solvent:

$$m = \frac{\text{no. moles solute}}{\text{no. kg solvent}}$$

In this case, we are given 26.8 g = 0.0268 kg of water, so:

$$m = \frac{\text{no. moles solute}}{0.0268}$$

Summarizing the analysis of the problem to this stage, we have:

$$1.40°C = 1.86°C \times \frac{\text{no. moles solute}}{0.0268}$$

To many students, it is by no means obvious what to do next. In order to complete the set-up of the problem, you must realize that the number of moles of solute can be set equal to the number of grams of solute (1.25 g) divided by the molecular weight (M).

$$\text{no. moles solute} = \frac{\text{no. grams solute}}{M} = \frac{1.25}{M}$$

(Notice again how essential it is that you be intimately acquainted with the mole concept.) Making the above substitution completes the set-up of the problem.

$$1.40°C = \frac{1.86°C \times 1.25}{M \times 0.0268}$$

Using simple algebra and arithmetic, this equation can be solved to give $M = 62.0$.

Whenever you use an algebraic equation to express a chemical principle, *you must be aware of the limitations of that equation*. To illustrate this point, consider the equation used in Example 1.3:

$$\Delta T_f = (1.86°C)m$$

This equation applies only to *water solutions*; it cannot be used to analyze freezing point depression problems in which a solvent other than water is involved. You could not, for example, use this equation to solve Problem 1.3e at the end of this chapter, where the solvent is benzene. The more general relationship is:

$$\Delta T_f = k_f m$$

where k_f is a proportionality constant whose numerical value depends upon the solvent ($1.86°C$ for water, $5.10°C$ for benzene, etc.).

Probably the most difficult type of problem in general chemistry is one whose analysis requires that you apply two or more principles. Example 1.4 illustrates a relatively simple case of this type.

Example 1.4 A sample of hydrogen gas is collected over water at 25°C (vapor pressure of water = 24 mm Hg). The total pressure of the wet gas is 752 mm Hg; its volume is 19.6 liters. Calculate the number of moles of hydrogen in the sample.

Solution Analysis of this problem requires that you be familiar with two principles concerning the physical behavior of gases.

1. The *Ideal Gas Law*, which relates the number of moles of a gas (n) to the volume that it occupies (V) at a particular pressure (P) and absolute temperature (T).

$$PV = nRT$$

where R, the so-called gas constant, is numerically equal to 62.3 when V is expressed in liters, P in mm Hg, and T in °K.

In order to apply this equation here, you must realize that to calculate the number of moles of hydrogen (n_{H_2}), the partial pressure of that gas (P_{H_2}) should be used rather than the total pressure (P_{tot}). To find the partial pressure of hydrogen, you use a second principle.

2. *Dalton's Law*, which states that the total pressure of a gas mixture is equal to the sum of the partial pressures of the individual components. In this case:

$$P_{tot} = P_{H_2} + P_{water}$$

The partial pressure of water, P_{water}, is equal to its vapor pressure at 25°C, 24 mm Hg. Substituting and solving for P_{H_2}, we obtain:

$$P_{H_2} = P_{tot} - P_{water} = (752 - 24) \text{ mm Hg} = 728 \text{ mm Hg}$$

Now, we can readily obtain the number of moles of hydrogen from the Ideal Gas Law.

$$n_{H_2} = \frac{P_{H_2} V}{RT} = \frac{(728 \text{ mm Hg})(19.6 \text{ liters})}{\left(62.3 \dfrac{\text{mm Hg liter}}{\text{mole }^\circ K}\right)(298^\circ K)} = 0.768 \text{ mole}$$

Curiously enough, students rarely have difficulty with this particular problem, perhaps because the two principles involved (Ideal Gas Law, Dalton's Law) are considered in the same chapter of the text and usually discussed in the same lecture or discussion. Like love and marriage, they follow one another in a logical sequence (regardless of which comes first). When the different principles required to analyze a problem are more separated in space and time, student performance drops off (Example 1.5).

Example 1.5 Analysis of a certain nonelectrolyte shows that it contains 38.7% by weight of carbon, 9.7% hydrogen, and 51.6% oxygen. A solution of 1.25 grams of this compound in 26.8 grams of water freezes at $-1.40^\circ C$. Use this information to obtain the molecular formula of the nonelectrolyte.

Solution In order to solve this problem, you must recognize that:
1. *The simplest formula, which gives the simplest atom ratio in the compound, can be obtained from the percentages by weight of the elements.* We will not take space here to illustrate how this is done; you may wish to refer to the appropriate chapter of your text or to Chapter 3 of this book for the principle involved. Suffice it to say that in this case the percentages given for carbon, hydrogen, and oxygen lead to the simplest formula, CH_3O.
2. *The molecular weight of the compound can be obtained from freezing point data as discussed in Example 1.3.* Noting that the data given here are identical with those in that example, we conclude that the molecular weight must again be 62.0.
3. *The molecular formula is readily obtained from the simplest formula and the molecular weight.* The formula weight of CH_3O is 31.0 (A.W. C + 3 × A.W. H + A.W. O = 12.0 + 3.0 + 16.0 = 31.0). Since the molecular weight, 62.0, is twice the formula weight of CH_3O, the molecular formula must be twice the simplest formula, or $C_2H_6O_2$.

It is unlikely that you could get very far with the analysis of Example 1.5 unless you had all three principles clearly in mind. The difficulty is that they are not ordinarily discussed together in the general chemistry course. The first and third principles are usually introduced early, perhaps in the first few weeks of the fall term. Principle 2, which deals with the calculation of molecular weight from freezing point data, usually comes much later. Sadly enough, by that time, many students have forgotten how to obtain simplest formulas (Principle 1) or how to relate them to molecular formulas (Principle 3). Like it or not, chemistry is a cumulative subject, where one principle builds upon another. You simply cannot afford to wipe your mind clean every time you take an examination.

The chances are very good that whenever you come across a problem that baffles you, more than one principle is involved. An effective way to analyze such a problem is to perform a sort of "reverse analysis," working backwards from the required quantity to ones that you know or can obtain from the information given in the problem. In Table 1.1, we show such an analysis applied to Example 1.5.

4. After you have analyzed the problem, set it up in a concise, logical manner.
Figure 1.3 shows how two students, whom we shall call A and F, worked out

Table 1.1 "Reverse Analysis" of a Multi-Principle Problem

Example 1.5

Reasoning:	(3) I could get the molecular formula if I knew the simplest formula and the molecular weight.
	(2) I can get the molecular weight from the freezing point data.
	(1) I can get the simplest formula from the percentages of C, H, and O (at least I hope I can). So I can solve the problem.

A solution of 1.25 grams of a certain nonelectrolyte in 26.8 grams of water freezes at −1.40°C. What is the molecular weight of the nonelectrolyte?

STUDENT A

$\Delta T_f = (1.86°C)m$

$\Delta T_f = 1.40°C$

$m = \frac{\text{no. moles solute}}{\text{no. kg solvent}} = \frac{1.25}{0.0268\ M}$

$\therefore 1.40°C = 1.86°C \times \frac{1.25}{0.0268\ M}$

$M = \frac{1.86°C \times 1.25}{1.40°C \times 0.0268} = 62.0$

Answer 62.0

STUDENT F

$\frac{1.40}{1.86} = 0.753$ $\frac{26.8}{1.25} = 21.4$ $T = 1.86\ m$

freezing point lowering

$.753 \overline{)\,.0466}$ .0620 (0.006)

$1.86 \overline{)\,21.4}$ ~~115~~ $M \times 1.25$

$\frac{1.25}{26.8}$ ~~.0466~~ $\frac{1.25}{M(.0268)}$

Answer see work

FIGURE 1.3 Two ways of setting up example 1.3.

the problem posed in Example 1.3. Student A is to be commended for indicating each step in the solution clearly and logically; her line of reasoning is clear at each stage. In contrast, student F seems to have followed a shotgun approach; he has made several false starts and superfluous calculations. An instructor assigned to grade this problem would find it impossible to follow his reasoning. If you search very carefully, you may locate the correct answer hidden among the scribbling,* but it is hard to say where it came from. Would you care to guess how much credit Student F received on this problem?

This is perhaps an extreme example of the contrast between the orderly and chaotic approaches to problem solving. Nevertheless, set-ups similar to that of Student F show up all too frequently on general chemistry examinations. Not only do they confuse the instructor, but more important, they suggest that the student himself is confused. Almost certainly, Student F did not really understand what he was doing. Under the mistaken impression that it would save time, he attempted to bypass the analysis of the problem, proceeding immediately to juggle numbers.

Occasionally, confronted with a several-step problem, a student analyzes it correctly and then, halfway through, forgets where he is going. To avoid disasters of this sort, many students find it helpful to write notes to themselves to serve as a reminder of the path they are following. Such notes act as a route map to guide you to your destination (Figure 1.4). They need not be elaborate or even intelligible to an outsider as long as you understand what they mean.

Problem: Example 1.5, p. 8

Notes	Translation
1. $\% \rightarrow S.F.$	1. Calculate the simplest formula from the percentages of C, O, and H.
2. $M: \Delta T_f = 1.86\,m$	2. Obtain the molecular weight, M, by using the equation for freezing point lowering.
3. $M + S.F. \rightarrow M.F.$ $M.F. = n\,(S.F.)$	3. Combine the simplest formula and the molecular weight to get the molecular formula. Remember that the molecular formula is an integral multiple of the simplest formula, i.e., n = 1, 2, – .

FIGURE 1.4 "Route maps" in problem solving.

5. Whenever possible, check to be sure that your answer makes sense.

A mistake, either in reasoning or arithmetic, can lead to an absurd answer to a problem. A moment's reflection on the magnitude of your answer may save you considerable embarrassment and perhaps suggest as well where you have gone wrong. Consider the problem listed in Figure 1.5, where at least 2 of 4 students came up with answers that fail to make sense. Can you tell, without reading on, which ones they are?

---

*This is known in the trade as the "multiple choice answer." Presumably, we are supposed to choose the correct answer from among the several possibilities.

Question: The equilibrium constant for the reversible reaction:

$$H_2(g) + I_2(g) \rightleftharpoons 2HI(g)$$

is 4.0. If one starts with 1.00 mole of H_2 and 0.100 mole of I_2 in a ten-liter container, how many moles of HI are produced at equilibrium?

Student Answers: (1) 0.125   (2) 1.49   (3) 0.182   (4) −3.50

FIGURE 1.5 Which are the *incorrect* answers?

Answer (4) is clearly ridiculous; it is impossible to have a negative number of moles. The other nonsensical answer is not quite as easy to spot. However, if you think about it, you will realize that (2) cannot be correct. With only 0.100 mole of I_2 available, no more than 0.200 mole of HI could be formed, even if the reaction went to completion (one mole of I_2 produces two moles of HI). The two remaining answers, 0.125 and 0.182, are both plausible since they fall between the limits 0.000 to 0.200, corresponding respectively to no reaction and complete reaction. If you actually set up the problem and solve for the number of moles of HI (this type of problem is discussed in the section of your text dealing with gaseous equilibria), you should find that the correct answer is 0.182.

Ordinarily, in order to detect an absurd answer, you must have a clear understanding of the terms and principles involved. This point is illustrated in Figure 1.6. Without carrying out any calculations, can you see why the answers quoted must be wrong?

Question 1:   A sample of gas occupies a volume of 120 cc at 20°C. If the temperature is increased to 50°C, what will the new volume be?

Student Answer: 106 cc

Question 2:   How much does a molecule of O_2 weigh?

Student Answer: 32 grams

FIGURE 1.6 Why *must* these answers be wrong?

In order to rule out the answer to Question 1 of Figure 1.6, you must be aware of the general principle that the volume of a gas is directly related to its temperature. An increase in temperature, from 20°C to 50°C, must increase the volume rather than decrease it, as an answer of 106 cc would imply. The answer to Question 2 may seem entirely reasonable to a student who has no "feeling" for the mass of a molecule. Only if you realize that molecules are tiny, invisible particles, far too small to be weighed on any balance, will you see that an O_2 molecule couldn't possibly weigh anything like 32 grams. As a matter of fact, it takes about 6×10^{23} molecules of O_2 to weigh 32 grams; if one could isolate a single molecule of oxygen and weigh it, it would tip the scales at about 5.3×10^{-23} gram (i.e., 0.000 000 000 000 000 000 000 053 gram).

Finally, we should point out that occasionally, in solving a problem, you will obtain a correct answer which surprises you, perhaps to the extent that you will be tempted to discard it and start all over again. The following example is a case in point.

Example 1.6 Determine the oxidation number of carbon in the acetic acid molecule, CH_3COOH.

Solution To solve this problem, you make use of the arbitrary rules by which oxidation numbers are assigned. One of these states that the sum of the oxidation numbers of all the atoms in a molecule is zero. Another says that, in compounds of this type, hydrogen has an oxidation number of +1, oxygen an oxidation number of −2. Applying these rules to acetic acid (2 carbon, 4 hydrogen, and 2 oxygen atoms) leads to the following simple algebraic equation:

$$0 = 2X + 4(+1) + 2(-2); \quad X = \text{oxid. no. of carbon}$$

Solving: $2X = 0 - 4 + 4 = 0$

$$X = 0 = \text{oxid. no. of carbon in } CH_3COOH$$

Most students would be at least a little bit shaken up by this answer. We tend to associate an oxidation number of zero with an elementary substance; at least, at first glance, it seems surprising to find an oxidation number of zero for an element in a compound substance. Yet, a careful check of the reasoning and the arithmetic should convince you that the answer is indeed correct. As a matter of fact, when you think about it, there is no reason why it shouldn't be. Since oxidation number is a man-made concept, established on the basis of arbitrary rules, an oxidation number of zero for carbon in acetic acid is as reasonable as any other.

Hopefully, Example 1.6 will convince you of the folly of changing your answer to a problem just because it "looks wrong." If common sense tells you that your answer is unusual, check your set-up, looking particularly for an error in reasoning. If no such error can be found, have enough confidence in yourself to leave things as they are.

1.2 COMPLICATIONS: TOO LITTLE OR TOO MUCH INFORMATION

Sometimes, when you analyze a problem in general chemistry, it appears that you do not have sufficient information to solve it. Conversely, you may be puzzled by what seems to be an oversupply of information, some of it unrelated to the problem at hand. Let's consider these two situations separately.

Too Little Data. Certain problems require assumptions which may be a source of concern to you. For instance, consider Question 1 in Figure 1.6, p. 11. Since the volume of a gas depends upon pressure as well as temperature, you may wonder why pressure is not mentioned in the statement of the problem. To work the problem with the information given, you have to assume that the pressure remains constant when the temperature increases. It would have been helpful if this condition had been specified, but instructors (and textbook writers) sometimes fail to do so.

Another case of a problem with a hidden assumption is Example 1.1, p. 4. Can you see what this assumption is? The statement of the problem neglects to mention how many moles of NO are available. The analysis that we went through assumes, in effect, that there is an excess of NO so that the yield of NO_2 depends only on the number of moles of O_2 available.

Assumptions such as those just described seldom bother students in general chemistry. Indeed, in these two cases, they may not even realize that they are making an assumption when they set up the problem. In contrast, most students realize that there is a vital piece of information missing in the statement of the following problem.

A sample of hydrogen gas is collected over water at 25°C. The total pressure of the wet gas is 752 mm Hg; its volume is 19.6 liters. Calculate the number of moles of hydrogen in the sample.

If you can't see at a glance what the missing piece of data is here, look back to Example 1.4, p. 7. Clearly, to solve this problem, you first have to obtain the partial pressure of hydrogen, which requires that you know *the vapor pressure of water at 25°C.*

Let's suppose for the sake of argument that you were faced with this problem on a homework assignment or an examination. What would you do? Logically, you would look for a table giving the vapor pressure of water as a function of temperature. Such a table, which might appear as an appendix in your textbook (use the index) or on the first page of the examination, would tell you that the vapor pressure of water at 25°C is 24 mm Hg. With this information, the analysis of the problem is straightforward (see Example 1.4).

Another, slightly different, example, taken directly from a general chemistry examination:

What is the concentration of OH⁻ ions in a 0.010 molar solution of HCl?

This question is completely baffling unless you realize that in any water solution, the concentrations of OH^- and H^+ are related by the equation:

$$(\text{conc } OH^-)(\text{conc } H^+) = 1.0 \times 10^{-14}$$

With this information, you should have little difficulty deciding that the concentration of OH^- is 1.0×10^{-12} molar (if you do, better read Chapter 4 on the use of exponents). You will find the equation just quoted in your textbook; it is used so frequently that your instructor may well expect you to know it and hence not include it on an examination paper.

Sometimes, a piece of missing information which appears to be vital at first glance turns out to be unnecessary. How would you answer this problem with no other information available?

Write a balanced net ionic equation for the reaction of benzene sulfonic acid, a strong acid, with OH⁻ ions.

Many students suppose that in order to write this equation, they must know the formula of benzene sulfonic acid. Not so! All you need do is apply the general principle that the reaction of *any* strong acid with OH^- ions is represented by the simple neutralization equation:

$$H^+(aq) + OH^-(aq) \rightarrow H_2O$$

In summary, before you give up on a problem because a vital piece of information seems to be missing:

(1) Make sure the information is really essential (remember the benzene sulfonic acid story).

(2) If you're sure you can't live without it, consider that it might be:

   (a) incorporated in a principle you're supposed to know (e.g., the relation between the concentrations of H^+ and OH^-).

   (b) available in a table you have access to (e.g., vapor pressure of water at $25°C$).

(3) If all else fails, ask your instructor about it. At the very least, he will be pleased that you have analyzed the problem to the point where you know something is missing. Who knows, he may even supply the data you need.

Too Much Data. To most students, there is nothing more exasperating than having more information than they need to work a problem. Just possibly, an instructor might be forgiven for the question in Example 1.7, but many students will complain that it was both unnecessary and confusing to specify the pressure.

Example 1.7 A sample of gas, confined at a constant pressure of 653 mm Hg, occupies a volume of 120 cc at 20°C. If the temperature is increased to 50°C, what will the new volume be?

Solution $V_2 = V_1 \times \dfrac{T_2}{T_1} = 120 \text{ cc} \times \dfrac{323°K}{293°K} = 132 \text{ cc}$

It would be a brave instructor indeed who dared to put Example 1.8, which reeks with excess information, on a general chemistry examination.

Example 1.8 Given the following information, calculate the heat of sublimation of ice in cal/mole.

    atomic weights: H = 1.0, O = 16.0   heat of vaporization water at

    specific heat ice: 0.50 cal/g °C   0°C: 600 cal/g

    specific heat water: 1.00 cal/g °C   melting point of ice: 0°C

    heat of fusion of ice: 80 cal/g   density of ice: 0.92 g/cc

Solution   heat of sublimation = heat of fusion + heat of vaporization

    per gram: heat of subl. = 80 cal/g + 600 cal/g = 680 cal/g

    per mole: heat of subl. = 680 cal/g $\times$ 18.0 g/mole = 12,200 cal/mole

---

Before you accuse your instructor of deliberately attempting to confuse you with this question, you would do well to reflect upon one of the facts of life: in practical problems, "excess information" is the rule rather than the exception. A doctor making a diagnosis must decide which of his patient's symptoms are pertinent and which are unrelated to his illness. A lawyer preparing a brief must search his law books to find precedents to back up his argument; in the process, he discards dozens of decisions that he feels are irrelevant to the case in hand. A chemist attempting to synthesize a new compound must choose, from the many methods reported in the literature, the particular one that he feels is most applicable to his task. Perhaps, therefore, it is not unreasonable to expect a student in general chemistry to develop some ability along these lines.

1.3 BAD HABITS TO AVOID

So far in this chapter, we have taken a positive approach to problem solving, recommending methods of analysis that should prove helpful to you. With some misgivings, we now consider the opposite side of the coin, warning you about techniques that are likely to do more harm than good. The difficulty is that the methods we are about to describe often exert a fatal fascination to beginning students who are unfamiliar with them.

Beware of Rote Methods. Some students will do almost anything to avoid analyzing problems. Instead, they rely on rote methods which supposedly allow you to solve a problem without understanding what you are doing. Such methods are deceptive in that they may seem to work well at first with very simple problems. Unfortunately, they let you down when you need them most. The first time you meet up with a problem a little out of the ordinary, the mechanical method leads you down the garden path to the wrong answer.

Every instructor has a "favorite" rote method which he warns his students against. Ours happens to be the so-called "ratio-and-proportion" method. We won't attempt to describe it here, in hopes you have never heard of it. If this is the case, please skip the following discussion. If, on the other hand, you are addicted to "ratios and proportions," the examples cited below may help you to kick the habit. We start with a relatively straightforward gas law problem, as it was "solved" by two students, one of whom understood what he was doing (Table 1.2).

Table 1.2 Student Responses to a Gas Law Problem

Problem: A flask, open to the atmosphere, contains 0.100 mole of gas at 20°C. If the flask is heated to 100°C (constant pressure and volume) how many moles of gas will remain in it?

	STUDENT A	STUDENT F
Analysis:	(1) $PV = nRT$	
	(2) P, V constant; n inversely proportional to T	"Zilch"
	(3) $n_2 = n_1 \times T_1/T_2$	
Solution:	$n_2 = 0.100 \times 293°\text{K}/373°\text{K}$	$\dfrac{0.100}{20} = \dfrac{x}{100}$
	$= 0.0786$	$x = \dfrac{0.100 \times 100}{20} = 0.500$

As you have probably deduced, the correct answer is the one obtained by Student A, 0.0786 mole. Student F, who depends upon the ratio-and-proportion approach, has assumed, without realizing it, that the number of moles is directly proportional to the temperature (in °C yet!) at constant pressure and volume. Indeed, anyone who uses this rote method is making that assumption, regardless of the particular problem. As it happens, most of the relationships discussed in general chemistry are *not* direct proportionalities. In this particular case, *n* is *inversely* related to *T*, as Student A has deduced by analysis of the problem.

We could cite many other examples of problems where the ratio-and-proportion approach leads to the wrong answer, but one more will suffice (Table 1.3).

The correct answer is, of course, 80.0 ml. As a matter of fact, Student F's answer is absurd; diluting 500 ml of one solution with water to give 200 ml of another solution would be a neat trick. However, students who rely on rote methods seldom worry about details like this.

Table 1.3   Student Responses to a Dilution Problem

Problem: How many ml of a 1.00 molar solution of NaCl should be diluted with water to give 200 ml of 0.400 molar NaCl?

	STUDENT A	STUDENT F
Analysis:	(1) no. moles solute unchanged on dilution	
	(2) no. moles = molarity × volume (liters)	
	(3) $V_1 \times 1.00 = (0.200 \text{ liter})(0.400)$	

Solution: $V_1 = 0.200 \text{ liter} \times \dfrac{0.400}{1.00}$

$= 0.0800 \text{ liter} = 80.0 \text{ ml}$

$\dfrac{x}{1.00} = \dfrac{200}{0.400}$

$x = \dfrac{200 \text{ ml} \times 1.00}{0.400} = 500 \text{ ml}$

Hopefully, these two case histories have convinced you of the shortcomings of mechanical methods of problem solving, of which "ratios and proportions" is but one example. *There is no substitute for analyzing a problem and understanding what you are doing.*

Go Easy on Analogies.   Most of the problems that appear on examinations in general chemistry are analogous to homework problems in the sense that they are based on the same chemical principles. Certainly the experience you gain in working problems outside of class will help you to analyze and solve other, similar problems on examinations. In this limited sense, "reasoning by analogy" can be a very useful technique.

However, you should be careful not to use this approach as an alternative to problem analysis. Too often, students attempt to solve a problem by mechanically repeating a set-up which, they seem to recall, gave them the correct answer in a similar situation. Table 1.4 shows two examples of misuse of the "reasoning by analogy"

Table 1.4   Misuse of the Reasoning by Analogy Approach

1. The mass of one molecule of a certain compound is 2.66×10^{-23} g. What is the mass of one mole of the compound (Avogadro's Number = 6.02×10^{23})?

	Student A	Student F
Reasoning:	One mole = Avog. no. of molecules	Seems to recall that in similar problem he had to divide by 6.02×10^{23}
Solution:	$6.02 \times 10^{23}(2.66 \times 10^{-23} \text{ g})$ $= 16.0 \text{ g}$	$\dfrac{2.66 \times 10^{-23} \text{ g}}{6.02 \times 10^{23}} = 0.442 \text{ g}^*$

2. The solubility of Ag_2CrO_4 is 1.0×10^{-4} mole/liter. Calculate K_{sp} of Ag_2CrO_4

$$Ag_2CrO_4 \text{ (s)} \rightleftharpoons 2Ag^+\text{(aq)} + CrO_4{}^{2-}\text{(aq)}$$

	Student A	Student F
Reasoning:	1) $K_{sp} = (\text{conc Ag}^+)^2(\text{conc CrO}_4{}^{2-})$ 2) conc $CrO_4{}^{2-} = 1.0 \times 10^{-4}$ conc $Ag^+ = 2.0 \times 10^{-4}$	Recalls that in similar problem involving AgCl, he squared the solubility to obtain K_{sp}.
Solution:	$K_{sp} = (2.0 \times 10^{-4})^2(1.0 \times 10^{-4})$ $= 4.0 \times 10^{-12}$	$K_{sp} = (1.0 \times 10^{-4})^2 = 1.0 \times 10^8$ *

*Apparently, Student F doesn't know how to use exponents either.

technique. Certainly you wouldn't follow Student F in the first case, but you might possibly fall into the same trap that he did in the second.

1.4 IF AT FIRST YOU DON'T SUCCEED . . .

Hopefully, the techniques of analysis suggested in this chapter will, if conscientiously followed, enable you to solve most of the problems encountered in general chemistry. However, we may as well be frank; sooner or later, you will come up against a problem that baffles you. All of us, from the beginning student to the most experienced instructor, face this situation from time to time. The question is: what should you do when you are stumped by a problem?

Our advice here depends upon the circumstances, specifically whether your hang-up comes on an examination or a homework assignment. If the problem that you can't analyze appears on an exam, we suggest that you forget it, at least temporarily. Try to push it into the back of your mind (a neat trick if you can do it), going on to answer the other questions on the examination. Only when you have done the best you can on the rest of the exam should you return to the problem that bothers you. Now, with a fresh viewpoint, you may be able to see how to analyze it. Quite possibly, in the process of working out solutions to other problems, a method of attack will occur to you. At any rate, the worst thing you can do is to get so bogged down on one problem that you have to hurry through the rest of the exam, making foolish mistakes that will exasperate you afterwards.

More frequently, your hang-up will come on a homework assignment, where you are applying a particular set of chemical principles for the first time. Here our advice is both simple and obvious: get help! When you are convinced that you have made an honest attempt to analyze the problem, consult with someone more experienced than you. This may be a fellow student,* your recitation instructor, or your lecturer. There are four suggestions that we would make in this area.

1. Try to carry your analysis of the problem as far as possible. In other words, try to find out what it is that bothers you. An explanation will make a great deal more sense if you have thought about the problem beforehand. This applies particularly to multi-principle problems, where you may be able to get half way to the answer on your own.

2. Don't be afraid to ask questions in class. Many students are reluctant to speak up, presumably because they are afraid of embarrassing themselves. As a matter of fact, if you can't understand a particular problem, the chances are that most of the class is having trouble with it. Someone has to break the ice, or your instructor won't be able to help you. That's what he's paid for; make sure you get your money's worth.

3. When your instructor is attempting to explain a problem, pay attention to what he is saying. Concentrate on the reasoning involved rather than the arithmetic, the analysis of the problem rather than its set-up. Don't waste time scribbling down the solution. After all, your objective is to learn how to apply the principles of chemistry to analyze a variety of problems, not to obtain a numerical answer to a specific problem.

---

*Remember, we're talking about *homework assignments*, not *examinations*!

4. After getting help on a problem, work an analogous one, this time completely on your own. Do this as soon as possible, while the principle involved is still fresh in your mind. Analogous problems, involving the same chemical principle, can ordinarily be found in your textbook or supplementary materials. If not, ask your instructor to make one up for you, or, better still, devise one of your own.

PROBLEMS

The problems listed here require only analysis; in no case are you asked for a numerical answer. However, you will need to be thoroughly familiar with the principles involved. It is suggested that you not attempt to answer the questions until you have studied the section of your text where the appropriate principles are discussed. The chapter references are to Masterton and Slowinski, "Chemical Principles," W. B. Saunders Company, 1973.

1.1 The following problems are taken from examination papers in general chemistry with one modification: certain numbers which appeared in the statement of the problem have been replaced by dashes to discourage number-juggling. In each case, you are to explain the meaning of each italicized term and show how an understanding of that term leads directly to the solution of the problem.

Example: Calculate the number of neutrons in a nucleus with *atomic number* 6 and *mass number* _____ .

Answer: atomic number = no. of protons; mass number = no. of protons + no. of neutrons. Hence, no. of neutrons = mass number – atomic number.

   a. A solid weighs 12.0 g; when added to a graduated cylinder containing _____ ml of liquid, the level rises to 15.0 ml. Calculate the *density* of the solid. (Chapter 1)
   b. The *atomic weight* of a certain element is _____ . How heavy, on the average, is an atom of this element compared to an atom of carbon-12? (Chapter 2)
   c. How many molecules of CO_2 are there in _____ *moles*? in _____ grams? (Chapter 3)
   d. When one gram of NiO is formed from the elements, $\Delta H =$ _____ kcal. What is the *heat of formation* of NiO? (Chapter 4)
   e. One gram of liquid water is added to an evacuated 250 ml flask; the equilibrium *vapor pressure* of water, _____ mm Hg, is established. If the flask had a volume of 125 ml instead of 250 ml, what pressure would be established? (Chapter 9)
   f. A solution is prepared by adding _____ moles of a certain solute to 112 grams of water, H_2O. What is the *molality* of the solute? (Chapter 10)

1.2 Each of the following problems can be solved using a single chemical principle. State that principle, which may take the form of an algebraic equation.

a. Write a balanced equation for the reaction of _____ with $O_2(g)$ to give $CO_2(g)$ and $H_2O(l)$. (Chapter 3)

b. Given a table of heats of formation, calculate ΔH in kcal for the reaction:

$$C_3H_8(g) + 5O_2(g) \rightarrow 3CO_2(g) + 4H_2O(l) \quad \text{(Chapter 4)}$$

c. What is the pressure in atmospheres exerted by _____ g of N_2 in a 10.0 liter container at $300°K$? (Chapter 5)

d. The vapor pressure of a certain liquid is 25.0 mm Hg at $20°C$ and _____ mm Hg at $50°C$. Estimate its heat of vaporization in cal/mole. (Chapter 9)

e. How many ml of $1.00\ M\ CaCl_2$ can be prepared by diluting 80 ml of _____ $M\ CaCl_2$ with water? (Chapter 10)

f. For a certain reaction, $\Delta H = +10.0$ kcal and $\Delta G =$ _____ kcal at $500°K$. Estimate ΔS for this reaction. (Chapter 12)

1.3 Each of the following problems requires more than one step for its solution. Indicate in some detail the path you would follow to solve the problem.

Example: Given the balanced equation: $2NO(g) + O_2(g) \rightarrow 2NO_2(g)$, calculate the number of grams of NO_2 that could be prepared from _____ moles of NO.

Path: (1) convert moles of NO to moles of NO_2 (2 moles NO $\doteqdot$ 2 moles NO_2)
(2) convert moles of NO_2 to grams of NO_2 (1 mole NO_2 = 46.0 g NO_2)

a. Calculate the number of atoms in _____ g of magnesium. (Chapter 2)

b. Combustion of 1.60 g of a certain hydrocarbon gives _____ g of H_2O. Determine the simplest formula of the hydrocarbon. (Chapter 3)

c. Calculate the average velocity in cm/sec of an O_2 molecule in a 250 ml flask containing one mole of O_2 in which the pressure of oxygen is _____ atm. (Chapter 5)

d. A certain metal crystallizes in a face-centered cubic structure. The atomic radius of the metal is _____ Å. What is the length of an edge of the unit cell? (Chapter 9)

e. A solution of 1.00 g of a certain compound in 12.0 g of benzene freezes at _____ °C. The freezing point of pure benzene is $5.50°C$ and its freezing point constant is $5.10°C$. Calculate the molecular weight of the solute. (Chapter 10)

1.4 Listed below are several equations commonly used in general chemistry. Explain clearly the meaning of each symbol in the equation.

a. $\Delta H = \Sigma \Delta H_f$ products $- \Sigma \Delta H_f$ reactants (Chapter 4)

b. $P_{tot} = P_1 + P_2 + \cdots$ (Chapter 5)

c. $PV = nRT$ (Chapter 5)

d. $\Delta T_b = k_b m$ (Chapter 10)

e. $\Delta G = \Delta H - T\Delta S$ (Chapter 12)

f. $\log X_0/X = kt/2.30$ (Chapter 14)

g. $K_b = 1.0 \times 10^{-14}/K_a$ (Chapter 17)

1.5 Below are the answers obtained by three different students for certain of the questions listed in Problems 1.2 and 1.3. Which of these answers must be wrong? Explain your reasoning.

	Question	Answer 1	Answer 2	Answer 3
a.	1.2b	0	+531	−531
b.	1.2c	2.0×10^{-5}	1.65	2.0×10^{5}
c.	1.2e	100	600	50
d.	1.3b	C_7H_8	C_2H_2	CH_3
e.	1.3c	5.0×10^{4}	5.0×10^{-4}	3.5×10^{4}
f.	1.3e	0.0100	100	63.0

1.6 In certain of the following set-ups, there is an error in reasoning. Explain the nature of the error and correct the set-up.

a. The element boron consists of two isotopes of masses 10.02 and 11.01; the abundance of the lighter isotope is 18.83%. Calculate the average atomic weight of boron. (Chapter 2)

Set-up: aver. A.W. $= \dfrac{10.02 + 11.01}{2} = 10.52$

b. A certain compound of cobalt and oxygen contains 73.4% by weight of Co. What is the simplest formula? (Chapter 3)

Set-up: atom ratio $Co/O = 73.4/26.6 = 2.76/1.00 \approx 11/4$

simplest formula: $Co_{11}O_4$

c. For the reaction: $N_2H_4(g) + O_2(g) \rightarrow N_2(g) + 2H_2O(l)$, $\Delta H = -148.7$ kcal, calculate the amount of heat evolved per gram of water formed. (Chapter 4)

Set-up: $\dfrac{148.7 \text{ kcal}}{36.0 \text{ g } H_2O} = 4.13 \text{ kcal/g } H_2O$

d. Calculate the volume occupied by 12.0 g of N_2 at 750 mm Hg and 27°C. (Chapter 5)

Set-up: $V = \dfrac{nRT}{P} = \dfrac{(12.0)(0.0821)(300)}{(28.0)(750)} = 0.0141$ liter

e. Calculate the freezing point of a solution containing 2.00 g of urea (molecular weight 60.0) in 20.0 g of water. (Chapter 10)

Set-up: $T = \dfrac{(1.86°C)(2.00)}{(60.0)(0.0200)} = 3.10°C$

f. For a certain reaction, ΔG at 300°K is +12.0 kcal; ΔH is +18.0 kcal. Calculate ΔG for this reaction at 600°K. (Chapter 12)

Set-up: $\Delta G = 18.0 \text{ kcal} - \dfrac{600}{300}(18.0 - 12.0) \text{ kcal}$

1.7 Certain of the following problems contain extra information not needed for the solution. In others, a necessary piece of information is missing. In each case, identify the extra data and/or indicate what additional data would be required to solve the problem.

a. A student calibrates a 20 ml pipet by filling it with pure water and allowing the water to run into a beaker which weighs 64.232 grams. He finds that the total mass (water + beaker) is 84.136 grams. What is the volume of the pipet? (Chapter 1)

b. Magnesium consists of three isotopes of masses 23.99, 24.99, and 25.99. It has an average atomic weight of 24.32. What are the relative abundances of the three isotopes? (Chapter 2)

c. Given the balanced equation: $4FeS(s) + 9O_2(g) + 4H_2O(l) \rightarrow 2Fe_2O_3(s) + 4H_2SO_4(l)$, calculate the number of grams of Fe_2O_3 that can be produced from 12.0 grams of FeS (A.W.: Fe = 55.8, S = 32.0, O = 16.0, H = 1.0). (Chapter 3)

d. The average velocity of an O_2 molecule at 25°C is 4.82×10^4 cm/sec. Calculate its average velocity at 100°C, assuming a constant pressure of one atmosphere ($R = 8.31 \times 10^7$ ergs/mole °K). (Chapter 5)

e. Calculate the pressure exerted by the water vapor in a 125 ml Erlenmeyer flask half-filled with water at 50°C (vapor pressure water = 92.5 mm Hg). (Chapter 9)

f. Calculate ΔG at 400°C for the reaction: $CO(g) + \frac{1}{2}O_2(g) \rightarrow CO_2(g)$, given that the free energies of formation of CO and CO_2 at 25°C are -32.8 and -94.3 kcal/mole respectively. (Chapter 12)

1.8 State problems for which the following set-ups would represent the correct solution. Be sure to include all the necessary information in your set-up.

Example: Set-up: $\quad 16.0 \text{ g}/6.02 \times 10^{23} = 2.66 \times 10^{-23}$ g

> *Problem:* Calculate the mass of a CH_4 molecule (Avogadro's no. = 6.02×10^{23}, A.W.:C = 12.0, H = 1.0)

a. *Set-up:* no. g CO_2 = 1.60 mole $CH_4 \times \dfrac{1 \text{ mole } CO_2}{1 \text{ mole } CH_4} \times \dfrac{44.0 \text{ g } CO_2}{1 \text{ mole } CO_2}$ (Chap. 3)

b. *Set-up:* $\Delta H = 2\Delta H_f H_2O(l)$ (Chapter 4)

c. *Set-up:* $\quad V_2 = 60.0 \text{ cc} \times \dfrac{719 \text{ mm Hg}}{760 \text{ mm Hg}} \times \dfrac{251°K}{294°K}$ (Chapter 5)

d. *Set-up:* molecular weight = $\dfrac{1.86 \times 1.60}{1.34 \times 0.0124}$ (Chapter 10)

e. *Set-up:* $K_{sp} = (1.3 \times 10^{-3})(2.6 \times 10^{-3})^2$ (Chapter 16)

f. *Set-up:* $K_b = \dfrac{1.0 \times 10^{-14}}{1.8 \times 10^{-5}}$ (Chapter 17)

CHAPTER 2

UNIT CONVERSIONS

In a sense, all problems in general chemistry involve a "conversion" from a given quantity to that for which you are asked to solve. Many times, little more than a change in units is involved, e.g., pounds to grams or grams to moles. Such exercises are readily solved by the "conversion factor" approach, which is the subject of this chapter. Indeed, as we shall see, this approach can be extended to solve rather more complex problems where more than one conversion is required.

To many beginning students, the conversion factor approach is new and hence, automatically, suspect. Applied to simple problems, it may seem awkward or artificial at first. As you gain more experience with the approach, you will become aware of two of its principal advantages. In the first place, it provides a convenient and quite general method of setting up solutions for a wide variety of problems in general chemistry. More important, it encourages you to analyze a problem to decide upon the path that you will follow to go from its statement to its solution.

2.1 SIMPLE CONVERSIONS

To illustrate the conversion factor approach, let us apply it to a particularly simple problem. Suppose we are asked to convert a length of 22 inches into feet. To do this, we make use of the conversion factor:

$$1 \text{ ft} = 12 \text{ in} \qquad (2.1)$$

Dividing both sides of this equation by *12 in* gives a quotient which is equal to unity:

$$\frac{1 \text{ ft}}{12 \text{ in}} = 1$$

If we now multiply *22 in* by this quotient, we do not change the value of the length, but we do accomplish the desired conversion of units:

$$22 \text{ in} \times \frac{1 \text{ ft}}{12 \text{ in}} = 1.8 \text{ ft*}$$

---

*In this and all other calculations, we follow the rules governing the use of significant figures (Chapter 7).

22

The conversion factor given by Equation 2.1 can be used equally well to convert a length given in feet, let us say 2.5 ft, to inches. In this case, we divide both sides of the equation by 1 ft to obtain:

$$\frac{12 \text{ in}}{1 \text{ ft}} = 1$$

Multiplying 2.5 ft by the ratio $\frac{12 \text{ in}}{1 \text{ ft}}$ converts the length from feet to inches:

$$2.5 \text{ ft} \times \frac{12 \text{ in}}{1 \text{ ft}} = 30 \text{ in}$$

Notice that a single conversion factor (e.g., 1 ft = 12 in) will always give us two quotients (1 ft/12 in or 12 in/1 ft) which are equal to unity. In making a conversion, we choose the quotient which will enable us to cancel out the unit that we wish to get rid of.

Reviewing the two calculations that we have just carried out, you may well object that all we have really done is to "divide by 12" in the first case and "multiply by 12" in the second. This is indeed true; with units such as feet and inches, which are thoroughly familiar to you, the conversion factor approach may seem unnecessary. However, when you are dealing with units that are less familiar, there is a distinct advantage in setting up the problem in a formal way, using the appropriate conversion factor. To illustrate this point, consider Example 2.1, where we are asked to convert a pressure expressed in atmospheres to pascals, the standard unit of pressure in the International System of Units (SI). Looking at the statement of the problem, it may not be entirely obvious to you whether you should "multiply" or "divide" by 1.01×10^5.

Example 2.1 Convert 2.36 atmospheres to pascals, using the conversion factor:

$$1 \text{ atm} = 1.01 \times 10^5 \text{ Pa}$$

Solution To apply the conversion factor approach here, we need a quotient in which pascals appear in the numerator and atmospheres in the denominator. This quotient is:

$$\frac{1.01 \times 10^5 \text{ Pa}}{1 \text{ atm}} = 1$$

Multiplying 2.36 atm by this quotient:

$$2.36 \text{ atm} \times \frac{1.01 \times 10^5 \text{ Pa}}{1 \text{ atm}} = 2.38 \times 10^5 \text{ Pa}$$

Here, as in all problems of this type, it is important that the units (e.g., pascals, atmospheres) be retained throughout the set-up of the problem. If the proper quotient (1.01×10^5 Pa/1 atm) is used, the original unit (atm) will cancel out to give the answer in the desired unit (Pa). If by mischance you get the conversion factor upside down (1 atm/1.01×10^5 Pa), your answer will come out in non-sensical units (atm^2/Pa), indicating a mistake in your reasoning.

2.2 MULTIPLE CONVERSIONS

Sometimes the conversion factor that is required to solve a problem is not available directly but must itself be calculated from other, simpler factors. As an example, suppose you were required to convert a volume of 2.12 in^3 to cm^3, given only the relationship:

$$1 \text{ in} = 2.54 \text{ cm} \tag{2.2}$$

Clearly, to carry out this conversion, you need a relation between *cubic* inches and *cubic* centimeters. This relation is readily obtained by cubing both sides of Equation 2.2.

$$(1 \text{ in})^3 = (2.54 \text{ cm})^3$$
$$1 \text{ in}^3 = 16.4 \text{ cm}^3 \tag{2.3}$$

Using the factor given by Equation 2.3, we find that 2.12 in^3 is equal to 34.8 cm^3.

$$2.12 \text{ in}^3 \times \frac{16.4 \text{ cm}^3}{1 \text{ in}^3} = 34.8 \text{ cm}^3$$

A different type of multistep conversion is illustrated in Example 2.2. Here, two different conversions are required: grams to pounds and cm^3 to in^3.

Example 2.2 The density of mercury is 13.6 g/cm^3. Express its density in lb/in^3, given that: 1 lb = 454 g; 1 in^3 = 16.4 cm^3.

Solution Let's proceed one step at a time, first converting grams to pounds and then cm^3 to in^3. To keep track of what we are doing, let's list the steps.
    (1) g → lb; (1 lb = 454 g)
    (2) cm^3 → in^3 ; (1 in^3 = 16.4 cm^3)
Now, the set-up:

$$\text{Density mercury} = \frac{13.6 \text{ g}}{cm^3} \times \frac{1 \text{ lb}}{454 \text{ g}} \times \frac{16.4 \text{ cm}^3}{1 \text{ in}^3} = 0.491 \frac{\text{lb}}{in^3}$$

Occasionally, you will be confronted with problems which require three or even four conversions. These can be handled exactly like the one just worked, setting up the various steps in succession. To avoid getting lost en route, it's a good idea to map out each step before you start.

Example 2.3 On a humid day in midsummer, the concentration of water vapor in the air is 0.0200 g/liter. If all of the water in a room with a volume of 612 ft^3 is removed by an air conditioner, how many pounds of water are condensed?

$$1 \text{ lb} = 454 \text{ g}; 1 \text{ ft}^3 = 28.3 \text{ liter}$$

Solution Reading this problem carefully, we see that our starting point is the volume of the room, 612 ft^3. We are supposed to find out how many pounds of water there are in that volume. A logical approach would be to follow a three-step path.
  (1) Convert ft^3 to liters; (1 ft^3 = 28.3 liter).
  (2) Convert liters to grams of water. According to the statement of the problem, on this particular day, 0.0200 grams of water are contained in one liter. Hence, our conversion factor is: 0.0200 g water $\doteq$ 1 liter, where the symbol $\doteq$ means "is equivalent to."
  (3) Convert grams to pounds of water; (1 lb = 454 g).
The suggested set-up is:

$$612 \text{ ft}^3 \times \frac{28.3 \text{ liter}}{1 \text{ ft}^3} \times \frac{0.0200 \text{ g water}}{1 \text{ liter}} \times \frac{1 \text{ lb}}{454 \text{ g}} = 0.763 \text{ lb water}$$
$$\quad\quad\quad (1) \quad\quad\quad\quad (2) \quad\quad\quad\quad (3)$$

Notice that so far as the set-up of the problem is concerned, the equivalence (2) is handled exactly like the equalities (1) and (3).

Hopefully, if you weren't a believer already, Example 2.3 will have convinced you of the merits of the conversion factor approach. We suggest you spring it on any of your friends who are still addicted to "ratios and proportions" and watch them struggle. As we shall see in the following section, many general chemistry problems are similar to Example 2.3 in that they involve several conversions.

2.3 APPLICATIONS IN GENERAL CHEMISTRY

One of the simplest types of conversions, which you will use over and over again in general chemistry, is that relating grams to moles or its converse, moles to grams (Example 2.4).

Example 2.4 Using a table of atomic weights, calculate:
  (a) the mass in grams of 0.825 mole of CO_2.
  (b) the number of moles in 6.24 grams of H_2O.

Solution

(a) To convert 0.825 mole of CO_2 to grams, we need the molecular weight of CO_2. This is readily found by summing the atomic weights of all the atoms in the formula.

$$\text{M.W. } CO_2 = \text{A.W. C} + 2(\text{A.W. O}) = 12.0 + 32.0 = 44.0$$

It follows that one mole of CO_2 weighs 44.0 grams, i.e.:

$$1 \text{ mole } CO_2 = 44.0 \text{ g } CO_2$$

Hence, the set-up becomes:

$$\text{no. g } CO_2 = 0.825 \text{ mole } CO_2 \times \frac{44.0 \text{ g } CO_2}{1 \text{ mole } CO_2} = 36.3 \text{ g } CO_2$$

(b) The molecular weight of water is 18.0 (i.e., $2 \times$ A.W. H + A.W. O = 2.0 + 16.0 = 18.0). In other words:

$$1 \text{ mole } H_2O = 18.0 \text{ g } H_2O$$

We use this conversion factor to go from grams of H_2O to moles:

$$\text{no. moles } H_2O = 6.24 \text{ g } H_2O \times \frac{1 \text{ mole } H_2O}{18.0 \text{ g } H_2O} = 0.347 \text{ mole } H_2O$$

Notice that the conversion factor relating moles to grams varies with the nature of the substance; 1 mole of CO_2 weighs 44.0 grams, 1 mole of H_2O only 18.0 grams. Every time we are required to convert moles to grams or vice-versa, we must first establish the appropriate conversion factor. This is readily done, given the formula of the substance and a table of atomic weights.

The conversion factor approach is particularly useful in solving problems dealing with balanced chemical equations. Example 2.5 illustrates the use of single and multiple conversion factors in a typical problem.

Example 2.5 The combustion of butane gas, C_4H_{10}, is represented by the balanced equation:

$$2C_4H_{10}(g) + 13O_2(g) \rightarrow 8CO_2(g) + 10H_2O(l).$$

Using this equation, calculate:

(a) the number of moles of CO_2 formed by the combustion of 1.42 moles of C_4H_{10}.

(b) the number of grams of H_2O formed from 1.42 moles of C_4H_{10}.

(c) the number of grams of O_2 required to form 21.0 grams of CO_2.

Solution

(a) The conversion factor required here is given directly by the co-efficients of the balanced equation. In this reaction, 2 moles of butane produce 8 moles of carbon dioxide. That is:

$$2 \text{ moles } C_4H_{10} \triangleq 8 \text{ moles } CO_2$$

Starting with 1.42 moles of C_4H_{10}, we must then have:

$$1.42 \text{ moles } C_4H_{10} \times \frac{8 \text{ moles } CO_2}{2 \text{ moles } C_4H_{10}} = 5.68 \text{ moles } CO_2$$

(b) Here we need to convert moles of C_4H_{10} to grams of H_2O. A logical two-step path would be:

(1) moles $C_4H_{10} \rightarrow$ moles H_2O; (2 moles $C_4H_{10} \triangleq 10$ moles H_2O)
(2) moles $H_2O \rightarrow$ grams H_2O; (1 mole H_2O = 18.0 g H_2O)
The set-up is:

$$\text{no. g } H_2O = 1.42 \text{ moles } C_4H_{10} \times \underbrace{\frac{10 \text{ moles } H_2O}{2 \text{ moles } C_4H_{10}}}_{(1)} \times \underbrace{\frac{18.0 \text{ g } H_2O}{1 \text{ mole } H_2O}}_{(2)}$$

$$= 128 \text{ g } H_2O$$

(c) Extending the analysis of parts (a) and (b), it would appear that a three-step path is in order:
(1) grams $CO_2 \rightarrow$ moles CO_2; (1 mole CO_2 = 44.0 g CO_2)
(2) moles $CO_2 \rightarrow$ moles O_2; (13 moles $O_2 \triangleq 8$ moles CO_2)
(3) moles $O_2 \rightarrow$ grams O_2; (1 mole O_2 = 32.0 g O_2)
Starting with 21.0 g of CO_2, the conversion is:

$$\text{no. g } O_2 = 21.0 \text{ g } CO_2 \times \underbrace{\frac{1 \text{ mole } CO_2}{44.0 \text{ g } CO_2}}_{(1)} \times \underbrace{\frac{13 \text{ moles } O_2}{8 \text{ moles } CO_2}}_{(2)} \times \underbrace{\frac{32.0 \text{ g } O_2}{1 \text{ mole } O_2}}_{(3)}$$

$$= 24.8 \text{ g } O_2$$

An alternative approach to multistep conversions of the type illustrated in parts (b) and (c) of Example 2.5 is to calculate in advance a single factor which will accomplish the required conversion directly. In part (b), we could have reasoned that since 2 moles $C_4H_{10} \triangleq 10$ moles H_2O, and one mole of water weighs 18.0 grams:

$$2 \text{ moles } C_4H_{10} \triangleq 10(18.0 \text{ g } H_2O) = 180 \text{ g } H_2O$$

Now we have precisely the conversion factor needed to go from 1.42 moles of C_4H_{10} to the number of grams of H_2O produced. The single-step conversion is:

$$1.42 \text{ moles } C_4H_{10} \times \frac{180 \text{ g } H_2O}{2 \text{ moles } C_4H_{10}} = 128 \text{ g } H_2O$$

Applying the same sort of reasoning in part (c), we start with the relationship given by the coefficients of the balanced equation

$$13 \text{ moles } O_2 \triangleq 8 \text{ moles } CO_2$$

and use the fact that one mole of O_2 weighs 32.0 grams while one mole of CO_2 weighs 44.0 grams to obtain directly the relationship we need, that between grams of O_2 and grams of CO_2.

$$13(32.0 \text{ g } O_2) \triangleq 8(44.0 \text{ g } CO_2)$$

$$416 \text{ g } O_2 \triangleq 352 \text{ g } CO_2$$

Now, with a single conversion, we find the number of grams of O_2 required to form 21.0 grams of CO_2:

$$21.0 \text{ g } CO_2 \times \frac{416 \text{ g } O_2}{352 \text{ g } CO_2} = 24.8 \text{ g } O_2$$

The approach described here leads, of course, to the same answer as that given under Example 2.5. Use whichever one appeals to you, i.e., whichever is easier for you to follow and apply.

Problems such as those discussed in Example 2.5 are ordinarily assigned early in the first term of the general chemistry course. Students sometimes forget that the same type of analysis can be applied in areas that are usually discussed much later in the course, such as electrochemistry (Example 2.6).

Example 2.6 Chrome plating is ordinarily carried out from an acidic solution of chromium(VI) oxide, CrO_3. The balanced half equation for the deposition of chromium metal is:

$$CrO_3(aq) + 6H^+(aq) + 6e^- \rightarrow Cr(s) + 3H_2O$$

How many grams of chromium can be plated by 20,500 coulombs (1 mole e^{-*} = 96,500 coul.)?

Solution We need a relation between coulombs and grams of chromium. The balanced equation gives us a relation between moles of electrons and moles of chromium:

$$6 \text{ moles } e^- \triangleq 1 \text{ mole } Cr$$

---

*One mole of electrons is often referred to as one *faraday*.

We can readily convert coulombs to moles of electrons using the conversion factor given in the statement of the problem. From there, we go to moles of chromium, using the equivalence written above. Finally, knowing the atomic weight of chromium, 52.0, we can readily convert moles to grams. In other words, a three-step process will take us from the quantity we are given, 20,500 coulombs, to the desired quantity, the number of grams of chromium plated.

(1) no. of coulombs → no. of moles of e⁻; (96,500 coulombs = 1 mole e⁻)

(2) no. of moles of e⁻ → no. of moles Cr; (6 moles e⁻ ≙ 1 mole Cr)

(3) no. of moles Cr → no. of grams Cr; (1 mole Cr = 52.0 g Cr)

$$\text{no. g Cr} = 20{,}500 \text{ coul.} \times \frac{1 \text{ mole e}^-}{96{,}500 \text{ coul.}} \times \frac{1 \text{ mole Cr}}{6 \text{ moles e}^-} \times \frac{52.0 \text{ g Cr}}{1 \text{ mole Cr}}$$

$$(1) \qquad\qquad (2) \qquad\qquad (3)$$

$$= 1.84 \text{ g Cr}$$

Notice the similarity between this example and part (c) of Example 2.5.

We could cite other examples of the usefulness of the conversion factor approach in general chemistry, but perhaps it is better to let you discover them for yourself. Instead, having promoted the advantages of the conversion factor method throughout this chapter, it is appropriate to end it with a couple of precautions.

(1) While the conversion factor approach is a valuable method of analyzing problems in general chemistry, it is by no means the only approach. Some students become so convinced of its merits that they try to apply it to every problem they encounter. Don't try to force a fit in this way. If, after reflection, the problem does not appear to lend itself to this approach, try another. It often turns out that a problem can be analyzed most simply in terms of an algebraic equation (Chapter 8) relating the quantity you are given to that for which you are asked to solve.

(3) Remember that the conversion factor approach is essentially a way of analyzing problems; it is not a "magic formula" for solving them. If you try to use it mechanically, it will become a rote method, no better than any other. We repeat here the theme of Chapter 1; in chemistry (and in just about everything else that we can think of), there is no substitute for understanding what you are doing.

PROBLEMS

Use the conversion factor approach to solve each of the following problems. If the necessary conversion factors are not given, look them up in your text or a handbook.

2.1 A crucible weighs 40.6 grams. Express its mass in
   a. kg   b. mg   c. lb   d. oz

2.2 Express the volume of a 125 ml Erlenmeyer flask in
  a. liters   b. cubic meters   c. quarts

2.3 On a certain day, the barometric pressure is given on a TV weather report as 30.10 inches of mercury. Convert this pressure to
  a. mm Hg   b. atm   c. bars (1 atm = 1.013 bar)

2.4 The density of bromine is 3.12 g/cm^3. Express its density in
  a. kg/m^3 b. lb/ft^3

2.5 An oxygen molecule at 25°C has an average velocity of 4.82 $\times$ 10^4 cm/sec. What is its velocity in miles/hr?

2.6 A sample of methane, CH_4, weighs 3.25 grams. How many moles of methane does this represent? How many molecules? (Avogadro's number = 6.02 $\times$ 10^{23}).

2.7 What is the mass in grams of
  a. 1.60 moles of $CaCl_2$?   b. 2.01 $\times$ 10^{23} molecules of H_2O?

2.8 Given the balanced equation:

$$4NH_3(g) + 5O_2(g) \rightarrow 4NO(g) + 6H_2O(l)$$

calculate
  a. The number of moles of O_2 required to react with 1.51 moles of NH_3.
  b. The number of grams of H_2O produced from 0.282 mole of NH_3.
  c. The number of moles of NH_3 required to form 6.40 grams of NO.
  d. The number of grams of NO formed from 9.80 grams of O_2.
  e. The number of kilograms of NH_3 required to form 1.00 lb of H_2O.

2.9 Consider the reaction:

$$2NH_3(g) \rightarrow N_2(g) + 3H_2(g)$$

Suppose one starts with 4.08 moles of NH_3 and forms x moles of N_2. Express, in terms of x:
  a. The number of moles of NH_3 left.
  b. The number of moles of H_2 formed.
  c. The number of grams of N_2 formed.

2.10 For the reaction: $CH_4(g) + 2O_2(g) \rightarrow CO_2(g) + 2H_2O(l)$, 213 kcal of heat are evolved per mole of CH_4 burned. Calculate the amount of heat evolved in kcal when
  a. 1.00 gram of CH_4 burns.
  b. 1.00 mole of O_2 is used.
  c. 1.00 lb of CO_2 is formed.
  d. a mixture of 8.00 g of CH_4 and 1.40 mole of O_2 is ignited. (Which reactant is in excess?)

2.11 The gas constant R has the value 0.0821 liter atm/mole °K. Calculate R in
  a. cal/mole °K (1 liter atm = 24.2 cal)
  b. ml mm Hg/mole °K

2.12 A quantum of UV light at a wavelength of 1000 Å has an energy of 1.99 X 10^{-11} erg. If each molecule of O_3 in a sample absorbs one quantum of this light, what is the absorption of energy in kcal/mole? (1 kcal = 4.18 X 10^{10} ergs; Avogadro's number = 6.02 X 10^{23}).

2.13 The balanced half equation for the reduction of PbO_2 in a storage battery is:

$$PbO_2(s) + 4H^+(aq) + SO_4{}^{2-}(aq) + 2e^- \rightarrow PbSO_4(s) + 2H_2O$$

Calculate the number of moles of PbO_2 reduced and the number of grams of $PbSO_4$ formed when 12,100 coulombs pass through this cell (96,500 coulombs = 1 mole of electrons).

2.14 An important property of a natural water supply is its biochemical oxygen demand (BOD) which is usually expressed in terms of the number of milligrams of oxygen required to react with the organic matter in one liter of water. A certain water supply contains 112 mg of ethyl alcohol per liter; it undergoes the following reaction with oxygen:

$$C_2H_5OH(aq) + O_2(g) \rightarrow CH_3COOH(aq) + H_2O$$

What is the BOD of this water?

2.15 In the nuclear fusion reaction: $2{}_1^2H \rightarrow {}_2^4He$, the mass of the product (one mole of helium) is 0.0256 grams less than that of the reactant (two moles of deuterium). The apparent loss in mass is exactly compensated for by the evolution of energy; 2.15 X 10^{10} kcal of energy are given off per gram of mass "lost." How much energy in kcal is given off when one gram of deuterium undergoes fusion?

CHAPTER 3

PER CENT

The concept of per cent is one that is familiar to all of us. A television commentator tells us that a certain candidate will receive 61 per cent of the total vote in a national election. In a somewhat different context, we read in the morning paper that the population of the earth is increasing at the rate of 2 per cent per year. Most of us are confident that we know what these numbers mean and how to use them in calculations. Just to make sure, the first section of this chapter reviews the arithmetic associated with percentages.

Many of the principles of chemistry are expressed in the language of per cent. The isotopic composition of elements, the proportions by weight of the elements in a compound, and the concentrations of the different components of a mixture are all conveniently stated as per cents. Terms such as "per cent dissociation" and "per cent yield" are used to describe the extent to which a chemical reaction takes place. In Section 3.2, we will discuss these chemical applications of per cent relationships.

3.1 BASIC RELATIONSHIPS

Per cent means literally "parts per hundred." When we say that a certain mixture of salt and sand contains 40% by weight of salt, we imply that in a 100 lb sample, there will be 40 lbs of salt. More generally, we mean that any sample, regardless of how much it weighs, will be 40/100 salt by weight. If, for instance, we were to select a 90 gram sample, it should contain:

$$\frac{40}{100} \times 90 \text{ g} = 36 \text{ g of salt}$$

In the general case of a mixture containing several components, one of which we will call A:

$$\frac{\% \text{ of } A}{100} \times \text{total amount} = \text{amount of } A$$

Rearranging this equation and solving for the per cent of A:

$$\% \text{ of } A = \frac{\text{amount of } A}{\text{total amount}} \times 100 \tag{3.1}$$

Two other useful relationships follow directly from Equation 3.1. One of these relates per cent to the decimal (or rational) fraction of a component in a mixture. Since the quotient, "amount of A/total amount," represents the fraction of A in the sample:

$$\% \text{ of } A = \text{fraction of } A \times 100 \tag{3.2}$$

The other expresses the fact that the sum of the percentages of all the components of a mixture must be 100. That is, for a mixture containing components A, B, $C \cdots$:

$$\% \text{ of } A + \% \text{ of } B + \% \text{ of } \dot{C} + \cdots = 100 \tag{3.3}$$

Example 3.1 Analysis of an iron-sulfur mixture weighing 12.50 grams shows it to contain 2.70 grams of iron. What are the percentages and decimal fractions of iron and sulfur?

Solution Let us first calculate the percentage of iron using Equation 3.1:

$$\% \text{ of iron} = \frac{\text{mass of iron}}{\text{total mass sample}} \times 100 = \frac{2.70 \text{ g}}{12.50 \text{ g}} \times 100 = 21.6\%$$

The percentage of sulfur follows directly (Equation 3.3):

$$\% \text{ of sulfur} = 100 - \% \text{ of iron} = 100 - 21.6 = 78.4$$

The decimal fractions of iron and sulfur are obtained by dividing the percentages by 100 (Equation 3.2):

$$\text{fraction of iron} = \frac{\% \text{ of iron}}{100} = \frac{21.6}{100} = 0.216$$

$$\text{fraction of sulfur} = \frac{\% \text{ of sulfur}}{100} = \frac{78.4}{100} = 0.784$$

Alternatively, in Example 3.1, the fraction of sulfur could have been obtained by subtracting the fraction of iron from 1:

$$\text{fraction of sulfur} = 1 - 0.216 = 0.784$$

In general:

$$\text{fraction } A + \text{fraction } B + \cdots = 1 \tag{3.4}$$

Frequently, the change in a quantity Q is expressed as a percentage of its initial value. We might, for example, describe the rate of inflation in the United States in 1973 by saying that the cost of living increased by 6 per cent. Used in this context, per cent change is defined by the relation:

$$\% \text{ change in } Q = \frac{\text{change in } Q}{\text{initial value of } Q} \times 100 = \text{fractional change in } Q \times 100 \tag{3.5}$$

Example 3.2 Liquid water expands by 4.2% when it is heated from 20°C to 100°C. What volume will be occupied at 100°C by a sample of water which has a volume of exactly one liter at 20°C?

Solution We use Equation 3.5 to calculate the increase in the volume of water above its initial value, V:

$$\text{change in } V = V \times \frac{\% \text{ change}}{100} = 1.000 \text{ liter} \times \frac{4.2}{100} = 0.042 \text{ liter}$$

The final volume at 100°C must then be:

$$1.000 \text{ liter} + 0.042 \text{ liter} = 1.042 \text{ liter}$$

3.2 APPLICATIONS IN CHEMISTRY

Many problems in general chemistry are phrased in terms of percentage composition or percentage change. In particular, there are four areas in the beginning course where the concept of per cent plays an especially important role. We shall now consider a few relatively straightforward examples, designed to illustrate the principles involved, in each of these areas. Additional problems, perhaps of a more advanced nature, can be found in the appropriate section of your textbook or supplementary material.

Concentrations of Solutions. The composition of a solution is often expressed in terms of the percentages of the various components. Here, we must note carefully the base upon which percentages are reported, i.e., the dimensions of the "total amount" and "amount of A" which appear in Equation 3.1. With liquid solutions, these quantities are perhaps most often expressed in terms of mass, in which case we refer to the **weight per cent** of a component.

$$\text{weight } \% \text{ of } A = \frac{\text{mass of } A}{\text{total mass solution}} \times 100 \tag{3.6}$$

Example 3.3 Concentrated hydrochloric acid contains 36% by weight of HCl.

    a. How many grams of HCl are there in 500 grams of this solution?

    b. What weight of this solution should you take to obtain 100 grams of HCl?

Solution

    a. Solving Equation 3.6 for the mass of HCl, we have:

$$\text{mass of HCl} = \text{total mass solution} \times \frac{\text{weight \% HCl}}{100}$$

$$= 500 \text{ g} \times \frac{36.0}{100} = 180 \text{ g}$$

    b. Here we require the total mass of solution:

$$\text{total mass solution} = \text{mass of HCl} \times \frac{100}{\text{weight \% HCl}}$$

$$= 100 \text{ g} \times \frac{100}{36.0} = 278 \text{ g}$$

Sometimes, particularly with gaseous solutions, we choose to express the concentrations of the components in terms of **mole per cent.*** The defining equation here is:

$$\text{mole \% } A = \frac{\text{no. moles } A}{\text{total no. moles}} \times 100 \qquad (3.7)$$

For a given component of a solution, the two quantities, weight per cent and mole per cent, will ordinarily differ significantly from one another, since the various components differ in molecular weight.

Example 3.4

    a. The weight per cent of alcohol, C_2H_6O, in a solution with water, H_2O, is 60.0. Calculate the mole per cent of alcohol.

    b. A certain brand of bottled gas is a mixture of propane, C_3H_8, and butane, C_4H_{10}. It contains 50 mole per cent of propane. What is the weight per cent of propane?

Solution

    a. Let us choose a fixed mass of solution to base our calculations upon; 100 grams is convenient. In 100 grams of solution, there are 60.0 grams of C_2H_6O and 40.0 grams of H_2O. Using a table of atomic weights, we find the molecular weights of C_2H_6O and

---

*For an *ideal, gaseous* solution, mole % = volume %

H_2O to be 46.0 and 18.0 respectively. Hence, in 100 grams of solution:

$$\text{no. moles } C_2H_6O = 60.0 \text{ g } C_2H_6O \times \frac{1 \text{ mole } C_2H_6O}{46.0 \text{ g } C_2H_6O}$$

$$= 1.30 \text{ moles } C_2H_6O$$

$$\text{no. moles } H_2O = 40.0 \text{ g } H_2O \times \frac{1 \text{ mole } H_2O}{18.0 \text{ g } H_2O} = 2.22 \text{ moles } H_2O$$

The total number of moles is: $1.30 + 2.22 = 3.52$

Hence, from Equation 3.7: mole % $C_2H_6O = \dfrac{1.30}{3.52} \times 100 = 36.9$

b. This time, we choose a fixed number of moles; let us say 100. In this quantity of gas, we have 50 moles of C_3H_8 and 50 moles of C_4H_{10}. The molecular weights of C_3H_8 and C_4H_{10} are 44.0 and 58.0 respectively. Hence, in 100 moles of gas:

$$\text{no. of grams } C_3H_8 = 50 \text{ moles } C_3H_8 \times \frac{44.0 \text{ g } C_3H_8}{1 \text{ mole } C_3H_8}$$

$$= 2200 \text{ g } C_3H_8$$

$$\text{no. of grams } C_4H_{10} = 50 \text{ moles } C_4H_{10} \times \frac{58.0 \text{ g } C_4H_{10}}{1 \text{ mole } C_4H_{10}}$$

$$= 2900 \text{ g } C_4H_{10}$$

The total mass is $2200 \text{ g} + 2900 \text{ g} = 5100 \text{ g}$

$$\text{weight % } C_3H_8 = \frac{2200 \text{ g}}{5100 \text{ g}} \times 100 = 43$$

Returning to Equation 3.6 for a moment, we note that the quotient "mass of A/ total mass of solution" represents the weight fraction of A in the solution. Again, in Equation 3.7, the quotient "no. moles A/total no. moles" is, by definition, the mole fraction of A. Thus, these equations might be written in a form analogous to Equation 3.2 as:

$$\text{weight % } A = \text{weight fraction } A \times 100 \qquad (3.8)$$

$$\text{mole % } A = \text{mole fraction } A \times 100 \qquad (3.9)$$

Mole fraction is a common and extremely useful concentration unit in general chemistry (Example 3.5).

Example 3.5 The mole per cents of N_2, O_2, and Ar in dry air are 78.1, 21.0, and 0.9, respectively.
  a. What are the mole fractions of these components?
  b. What is the molecular weight of air?

Solution
  a. We see from Equation 3.9 that mole fraction can be obtained by dividing mole per cent by 100. Hence, the mole fractions of N_2, O_2, and Ar must be 0.781, 0.210, and 0.009, respectively.
  b. Let us calculate the mass in grams of one mole of air. Noting that we have 0.781 mole of N_2 (M.W. = 28.0), 0.210 mole of O_2 (M.W. = 32.0), and 0.009 mole of Ar (M.W. = 39.9), we have:

$$0.781 \times 28.0 \text{ g} + 0.210 \times 32.0 \text{ g} + 0.009 \times 39.9 \text{ g}$$

$$= 21.9 \text{ g} \quad + \quad 6.7 \text{ g} \quad + \quad 0.4 \text{ g} \quad = 29.0 \text{ g}$$

We conclude that the molecular weight of air must be 29.0.

Isotopic Abundances. Most elements exist in nature as a homogeneous mixture of different isotopes, i.e., atoms of different masses. For example, chlorine occurs as a mixture of two isotopes, Cl-35 and Cl-37, which have atomic weights on the C-12 scale of 34.97 and 36.97, respectively. The relative abundances of the different isotopes of an element are usually specified by giving their mole per cents.* In the case of chlorine, the mole per cents of Cl-35 and Cl-37 are 75.77 and 24.23 respectively.

Knowing the atomic weights and abundances of the isotopes of an element, its atomic weight is readily calculated following essentially the same procedure used in Example 3.5(b).

Example 3.6 From the information just given, calculate the atomic weight of chlorine.

Solution We are asked in effect to calculate the mass in grams of one mole of Cl. To do this, we proceed as in Example 3.5(b), first writing down the mole fractions of the two components.

mole fraction Cl-35 = 75.77/100 = 0.7577; mole fraction Cl-37 = 0.2423

In one mole of Cl, we have 0.7577 mole of Cl-35, atomic weight 34.97, and 0.2423 mole of Cl-37, atomic weight 36.97. The mass of one mole of Cl must then be:

$$0.7577 \times 34.97 \text{ g} + 0.2423 \times 36.97 \text{ g} = 26.49 \text{ g} + 8.96 \text{ g} = 35.45 \text{ g}$$

The atomic weight of chlorine must then be 35.45.

*Usually referred to simply as "per cent."

For an element consisting of only two isotopes, it is possible to calculate the relative abundances of the isotopes from the known value of the atomic weight of the element (Example 3.7).

Example 3.7 Boron consists of two isotopes: B-10, atomic weight 10.013, and B-11, atomic weight 11.009. The atomic weight of the element itself is 10.811. What are the mole fractions and per cent abundances of the two isotopes?

Solution The fundamental relationship is:

A.W. B = mole fract. B-10(A.W. B-10) + mole fract. B-11(A.W. B-11)

Substituting the atomic weights given in the statement of the problem:

10.811 = (mole fract. B-10) $\times$ 10.013 + (mole fract. B-11) $\times$ 11.009

Let us represent by x the mole fraction of B-11. Since the two mole fractions must add to 1, the mole fraction of B-10 must be $1 - x$. Making this substitution:

$$10.811 = 10.013(1 - x) + 11.009x$$

Solving this equation for x:

10.811 - 10.013 = 11.009x - 10.013x;

0.798 = 0.996x;

$$x = \frac{0.798}{0.996} = 0.802$$

We conclude that the mole fractions of B-11 and B-10 are 0.802 and $(1 - 0.802) = 0.198$, respectively. The per cent abundances (mole %'s) are obtained by multiplying by 100:

$$\% \text{ abundance B-11} = 80.2; \text{B-10} = 19.8$$

Formulas of Compounds: Per Cent Composition. The Law of Constant Composition states that a pure compound always contains the same elements in the same proportions by weight. The "proportion by weight" of an element in a compound is ordinarily expressed by giving its weight per cent. The weight per cents of the various elements making up a compound are readily determined if its formula is known (Example 3.8).

Example 3.8 Calculate the percentages by weight of the elements in potassium chromate, K_2CrO_4.

Solution In one mole of K_2CrO_4, there must be 2 moles of K (78.2 g), 1 mole of Cr (52.0 g), and 4 moles of O (64.0 g). The mass of one mole of K_2CrO_4 is found by adding these three masses:

$$2 \text{ moles K} = 2 \times 39.1 \text{ g} = 78.2 \text{ g}$$
$$1 \text{ mole Cr} \qquad\qquad = 52.0 \text{ g}$$
$$4 \text{ moles O} = 4 \times 16.0 \text{ g} = \underline{64.0 \text{ g}}$$
$$1 \text{ mole } K_2CrO_4 \qquad = 194.2 \text{ g}$$

The weight percentages of the three elements follow:

$$\text{weight \% K} = \frac{78.2 \text{ g}}{194.2 \text{ g}} \times 100 = 40.3; \text{weight \% Cr} = \frac{52.0 \text{ g}}{194.2 \text{ g}} \times 100 = 26.8$$

$$\text{weight \% O} = 100.0 - 40.3 - 26.8 = 32.9$$

Since the weight percentages of the elements in a compound can be determined from its formula, the reverse operation should also be possible. That is, we should be able to obtain the formula of a compound knowing the percentages by weight of the elements. It is indeed possible to do this, although the analysis is somewhat less obvious than that illustrated in Example 3.8. The quantity obtained from such a calculation is the **simplest formula**, which gives the simplest whole number ratio between the numbers of atoms of the different elements making up the compound.

Example 3.9 The weight percentages of calcium, chlorine, and oxygen in a certain compound are 22.9, 40.5, and 36.6, respectively. What is its simplest formula?

Analysis
1. Let us first calculate the number of moles of each element in a fixed weight of the compound: for convenience, 100 g. This calculation is readily done if we know the atomic weights of the elements (A.W. Ca = 40.0, Cl = 35.5, O = 16.0).
2. The three numbers that we obtain in this way represent the relative numbers of moles of Ca, Cl, and O in any weight of the compound. *More important, they also represent the relative numbers of atoms of the three elements*, since the conversion factor relating moles to atoms (1 mole = 6.02×10^{23} atoms) is the same for all elements.
3. To obtain the simplest formula, we need only find the simplest whole number ratio between the three numbers referred to in (1) and (2).

Set-up

1. In 100 grams of the compound, there are 22.9 grams of Ca, 40.5 grams of Cl, and 36.6 grams of O. Hence, we have:

$$\text{no. moles Ca} = 22.9 \text{ g Ca} \times \frac{1 \text{ mole Ca}}{40.0 \text{ g Ca}} = 0.572 \text{ mole Ca}$$

$$\text{no. moles Cl} = 40.5 \text{ g Cl} \times \frac{1 \text{ mole Cl}}{35.5 \text{ g Cl}} = 1.14 \text{ mole Cl}$$

$$\text{no. moles O} = 36.6 \text{ g O} \times \frac{1 \text{ mole O}}{16.0 \text{ g O}} = 2.29 \text{ mole O}$$

2. The atom ratio is 0.572 atom Ca:1.14 atom Cl:2.29 atom O
3. One way to arrive at the simplest, whole-number atom ratio is to divide each of these numbers by the smallest, 0.572:

$$\text{Ca:} \frac{0.572}{0.572} = 1.00; \text{Cl:} \frac{1.14}{0.572} = 2.00; \text{O:} \frac{2.29}{0.572} = 4.00$$

The simplest formula must clearly be $CaCl_2O_4$.

Chemical Equations: Per Cent Dissociation, Per Cent Yield. In working problems based on balanced chemical equations, we often express the extent of reaction in terms of per cent. As an example, to indicate the degree to which 0.10 molar acetic acid dissociates into ions:

$$HC_2H_3O_2(aq) \rightarrow H^+(aq) + C_2H_3O_2^-(aq)$$

we might say that it is "1.3 per cent dissociated" (or 1.3 per cent ionized). In another case, we could describe the extent to which toluene is converted to TNT (trinitrotoluene):

$$C_7H_8(l) + 3HNO_3(l) \rightarrow C_7H_5(NO_2)_3(s) + 3H_2O(l)$$

by quoting a "per cent yield" of TNT of 75.0

Per Cent Dissociation The per cent dissociation of a reactant is defined as:

$$\% \text{ dissociation of } R = \frac{\text{amount of } R \text{ dissociated}}{\text{original amount of } R} \times 100 \qquad (3.10)$$

$$= \text{fraction of } R \text{ dissociated} \times 100$$

The application of Equation 3.10 to a simple problem is illustrated in Example 3.10.

Example 3.10 When 0.80 mole of acetic acid is added to a liter of water, 0.0038 mole of H^+ is produced by the reaction:

$$HC_2H_3O_2(aq) \rightarrow H^+(aq) + C_2H_3O_2^-(aq)$$

What is the % dissociation of acetic acid under these conditions?

Solution From the equation for the reaction, we see that the 0.0038 mole of H^+ must have come from the dissociation of an equal number of moles, 0.0038, of $HC_2H_3O_2$ (1 mole $HC_2H_3O_2 \rightarrow$ 1 mole H^+). Hence:

$$\% \text{ dissoc. } HC_2H_3O_2 = \frac{\text{amt. } HC_2H_3O_2 \text{ dissoc.}}{\text{orig. amt. } HC_2H_3O_2} \times 100$$

$$= \frac{0.0038}{0.80} \times 100 = 0.48\%$$

Per Cent Yield The per cent yield of a product in a reaction is defined as:

$$\% \text{ yield of } P = \frac{\text{actual yield of } P}{\text{theoretical yield of } P} \times 100 \qquad (3.11)$$

The theoretical yield is the amount of product that would be obtained by the complete conversion of reactants. In a particular reaction, the per cent yield can vary from 0 to 100, depending upon the conditions used, the skill of the person carrying out the reaction, and other factors. As you can imagine, we are ordinarily happiest when the per cent yield approaches the upper limit of 100. On the other hand, when a student reports a per cent yield greater than 100, we suspect his calculations, his technique, or his honesty (usually in that order).

Example 3.11 Consider the reaction of toluene with nitric acid to give TNT:

$$C_7H_8(l) + 3HNO_3(l) \rightarrow C_7H_5(NO_2)_3(s) + 3H_2O(l)$$

A student reacting 20.0 grams of toluene, C_7H_8, with excess nitric acid obtains 21.5 grams of TNT, $C_7H_5(NO_2)_3$. What is the per cent yield?

Solution Before we can use Equation 3.11, we must calculate the theoretical yield, i.e., the number of grams of TNT that would be produced by the complete conversion of 20.0 grams of toluene. To do this, we follow the three-step path:

    1. grams toluene $\rightarrow$ moles toluene (1 mole C_7H_8 = 92.0 grams)
    2. moles toluene $\rightarrow$ moles TNT (1 mole $C_7H_8 \rightarrow$ 1 mole TNT)
    3. moles TNT $\rightarrow$ grams TNT (1 mole $C_7H_5(NO_2)_3$ = 227 grams)

$$\text{theor. yield TNT} = 20.0 \text{ g } C_7H_8 \times \frac{1 \text{ mole } C_7H_8}{92.0 \text{ g } C_7H_8} \times \frac{1 \text{ mole TNT}}{1 \text{ mole } C_7H_8} \times \frac{227 \text{ g TNT}}{1 \text{ mole TNT}}$$

$$= 49.3 \text{ g TNT}$$

Now we can readily obtain the % yield:

$$\text{% yield TNT} = \frac{\text{actual yield TNT}}{\text{theor. yield TNT}} \times 100 = \frac{21.5 \text{ g}}{49.3 \text{ g}} \times 100 = 43.6$$

3.3 PARTS PER MILLION (PPM), PARTS PER BILLION (PPB)

As we have seen, it is often convenient to express the concentration of a component of a solution in per cent or "parts per hundred." There is, of course, nothing unique or sacred about "parts per hundred"; we could, if we wished, express concentration in "parts per thousand," or even "parts per 24,051." For trace components of a mixture, concentration is sometimes expressed in "parts per million" (ppm). Thus, we might describe the level of mercury in a contaminated sample of tunafish by saying that its concentration is "1.05 ppm." This description implies that if we analyzed a million grams of tunafish (heaven forbid), we would find 1.05 grams of mercury. More realistically, in a one gram sample, there should be 1.05/1,000,000 or 1.05×10^{-6} gram of mercury.* The general relationship for a component A of a mixture is:

$$\text{ppm of } A = \frac{\text{amount of } A}{\text{total amount sample}} \times 1,000,000 = \text{fraction of } A \times 10^6 \quad (3.12)$$

For very, very dilute solutions, concentrations are often expressed in "parts per billion" (ppb).

$$\text{ppb of } A = \frac{\text{amount of } A}{\text{total amount sample}} \times 1,000,000,000 = \text{fraction of } A \times 10^9 \quad (3.13)$$

The relationship between these two concentration units is readily obtained by dividing Equation 3.13 by 3.12:

$$\frac{\text{ppb of } A}{\text{ppm of } A} = \frac{10^9}{10^6} = 10^3 = 1000 \quad (3.14)$$

Example 3.12 The concentration of gold in sea water is estimated to be about 0.0040 ppb.
   a. Express its concentration in ppm.
   b. How much sea water would have to be extracted to obtain one gram of gold?

Solution

   a. From Equation 3.14:

$$\text{ppm of gold} = \frac{\text{ppb of gold}}{1000} = \frac{0.0040}{1000} = \frac{4.0 \times 10^{-3}}{10^3} = 4.0 \times 10^{-6}$$

---

*See Chapter 4 for a discussion of exponential numbers.

b. Using Equation 3.13:

$$\text{total amt. sea water} = \text{amt. of gold} \times \frac{10^9}{\text{ppb of gold}}$$

$$= 1.00 \text{ g} \times \frac{10^9}{0.0040} = \frac{1 \times 10^9 \text{ g}}{4.0 \times 10^{-3}} = 2.5 \times 10^{11} \text{ g}$$

(This is roughly equivalent to a cube of sea water 200 feet on an edge).

In expressing concentrations in ppm or ppb, one should be careful to specify whether they are calculated on a weight basis (e.g., grams per million grams, pounds per billion pounds) or on a mole basis (moles per 10^6 or 10^9 total moles). Unfortunately, this distinction is seldom made. In the absence of any information to the contrary, it is usually safe to assume that "ppm" or "ppb" refers to a weight basis for a solid or liquid mixture. In gaseous solutions, the convention is the opposite; ppm or ppb are ordinarily quoted on a mole basis. Thus, if we read that the concentration of sulfur dioxide in polluted air is 2.0 parts per million, we understand that there are 2.0 moles of SO_2 in one million moles of air or 2.0×10^{-6} mole of SO_2 per mole of air.

PROBLEMS

3.1 Analysis of a 1.435 gram sample of iron ore shows that it contains 0.369 gram of iron.
   a. What is the per cent of iron in the ore? the decimal fraction?
   b. How many pounds of ore must be taken to give one pound of iron?

3.2 A certain brand of household disinfectant contains 3.0% by weight of hydrogen peroxide and sells for 69 cents a pound. A competitive brand contains 4.6% hydrogen peroxide and sells for 89 cents a pound. If hydrogen peroxide is the only active ingredient in both cases, which is the better buy?

3.3 A certain brand of aspirin tablets is advertised as being 98% pure. How many grams of aspirin can be obtained from one pound of tablets?

3.4 The volume of a sample of gas maintained at constant pressure is 12.0 liters at $0°C$ and increases by 0.366% of this value per $°C$ when heated.
   a. What will be the volume of the sample at $100°C$?
   b. At what temperature will the gas have a volume twice that at $0°C$?

3.5 A salt solution contains 12.2% by weight of sodium chloride. If 450 grams of solution are evaporated to dryness, how much salt will be obtained? How much water must be evaporated?

3.6 Concentrated nitric acid, HNO_3, contains 72.0% by weight of HNO_3.
    a. How many moles of HNO_3 are in one kilogram of acid?
    b. How many moles of HNO_3 are in one liter of acid ($d = 1.42$ g/ml)?

3.7 Calculate the weight per cent and weight fraction of $CaCl_2$ in a 1.00 molal solution (molality = no. moles solute per kilogram of solvent).

3.8 A certain gas mixture of helium, He, and oxygen, O_2, contains 80.0 mole per cent of helium. What are the mole fractions of He and O_2? What is the molecular weight of the gas?

3.9 Using the data given in Example 3.5, calculate the weight per cents of N_2, O_2, and Ar in air.

3.10 Iron consists of four isotopes with the following masses and abundances:

mass	53.96	55.95	56.95	57.95
%	5.81	91.64	2.21	0.34

Calculate the atomic weight of iron.

3.11 The atomic weight of copper is 63.54; it consists of two isotopes, Cu-63 and Cu-65, with atomic weights of 62.96 and 64.96, respectively. Calculate the mole per cents of the two isotopes.

3.12 Calculate the percentages of the elements in:
    a. $CaCl_2$ b. H_3PO_4 c. $Co(NH_3)_3(NO_2)_3$

3.13 Determine the simplest formulas of compounds containing:
    a. 19.3% Na, 26.8% S, 53.9% O   b. 19.2% P, 2.5% H, 78.3% I

3.14 A solution prepared by dissolving 0.100 mole of acetic acid to give one liter of solution is found to be 1.3% dissociated via the reaction:

$$HC_2H_3O_2(aq) \rightarrow H^+(aq) + C_2H_3O_2^-(aq)$$

What are the concentrations, in moles/liter, of H^+, $C_2H_3O_2^-$, and $HC_2H_3O_2$?

3.15 At a certain temperature and pressure, NO_2 is partially dissociated via the reaction:

$$2NO_2(g) \rightarrow 2NO(g) + O_2(g)$$

When 2.16 mole of NO_2 is allowed to reach equilibrium, 0.26 mole of O_2 is formed.
    a. What is the % dissociation of NO_2 under these conditions?
    b. What are the mole fractions of NO_2, NO, and O_2 in the equilibrium mixture?

3.16 In the reaction:

$$C_2H_4(g) + H_2O(l) \rightarrow C_2H_5OH(l)$$

24.0 grams of C_2H_4 is reacted with excess water to produce 16.0 grams of C_2H_5OH. What is the per cent yield?

3.17 A student who needs 65 grams of hydrogen bromide is advised to prepare it by heating sodium bromide with an excess of phosphoric acid:

$$NaBr(s) + H_3PO_4(l) \rightarrow NaH_2PO_4(s) + HBr(g)$$

He is told that he can expect a yield of only about 35%. How much NaBr should he start with?

3.18 The solubility of DDT in water is estimated to be about 0.00005 g/100 g of water. Express the solubility in:
   a. weight per cent   b. ppm   c. ppb

3.19 The concentration of neon in air is 18.2 ppm. Under normal conditions, a mole of air occupies about 25 liters.
   a. What is the concentration of neon in moles/liter?
   b. How many liters of air would have to be processed to extract one gram of neon (A.W. = 20.2)?

CHAPTER 4

EXPONENTIAL NUMBERS

In chemistry, we frequently deal with very large or very small numbers. In one gram of the element carbon there are:

$$50,150,000,000,000,000,000,000$$

atoms of carbon. At the opposite extreme, the mass of a single carbon atom is:

$$0.0000000000000000000001994 \text{ gram}$$

Numbers such as these are not only difficult to write, they are very awkward to work with. Imagine how tedious it would be to find the mass of 2150 carbon atoms by carrying out the operation:

$$\frac{0.0000000000000000000001994 \text{ gram}}{\times 2150}$$

One way to simplify operations of this type is to use what we call **exponential notation**. In exponential notation, numbers such as those written above are expressed as a number between one and ten (**coefficient**) times an integral power of ten (**exponential**). Examples of exponential numbers include:

$$1 \times 10^4; \quad 2.23 \times 10^3; \quad 5.6 \times 10^{-4}$$

To understand precisely what these exponential numbers mean, it may be helpful to refer to Table 4.1.

For the three numbers cited previously, we have:

$$1 \times 10^4 = 1(10,000) = 10,000$$

$$2.23 \times 10^3 = 2.23(1000) = 2230$$

$$5.6 \times 10^{-4} = 5.6(0.0001) = 0.00056$$

Table 4.1 Exponentials

$$10^6 = (10)(10)(10)(10)(10)(10) = 1,000,000$$
$$10^5 = (10)(10)(10)(10)(10) \quad = \quad 100,000$$
$$10^4 = (10)(10)(10)(10) \quad = \quad 10,000$$
$$10^3 = (10)(10)(10) \quad = \quad 1000$$
$$10^2 = (10)(10) \quad = \quad 100$$
$$10^1 = (10) \quad = \quad 10$$
$$10^0 = \quad = \quad 1$$
$$10^{-1} = (0.1) \quad = \quad 0.1$$
$$10^{-2} = (0.1)(0.1) \quad = \quad 0.01$$
$$10^{-3} = (0.1)(0.1)(0.1) \quad = \quad 0.001$$
$$10^{-4} = (0.1)(0.1)(0.1)(0.1) \quad = \quad 0.0001$$

4.1 WRITING NUMBERS IN EXPONENTIAL NOTATION

In order to make use of exponential notation, we must be able to write any number, large or small, as an exponential number. To understand how this is done, it may be helpful to start with two relatively simple cases which can be worked directly from the entries in Table 4.1.

Suppose we wish to express the number 5196 in exponential notation. We realize that this number can be written as:

$$5.196 \times 1000$$

Referring to Table 1.1, we note that $1000 = 10^3$. Therefore:

$$5196 = 5.196 \times 1000 = 5.196 \times 10^3$$

As another illustration, consider the number 0.0028. To express this number in exponential notation, we first write it as:

$$2.8 \times 0.001$$

Since $0.001 = 10^{-3}$, we have:

$$0.0028 = 2.8 \times 0.001 = 2.8 \times 10^{-3}$$

The method which we used to work these two examples is not particularly useful with extremely large or extremely small numbers, for which tables of exponentials are seldom available. We can, however, deduce from these examples a more general approach to the problem. Notice that when we expressed 5196 in exponential notation, we arrived at an exponent of 3; this is the *number of places which the decimal point must be moved (to the left) to give the coefficient, 5.196*. Again, in expressing 0.0028 in exponential notation, the exponent, -3, is the *number of places which the decimal point must be moved (to the right) to give the coefficient, 2.8*. In general:

To express a number in exponential notation, write it in the form:

$$C \times 10^n$$

where C is a number between 1 and 10 (e.g., 1, 2.62, 5.8) and n is a positive or negative integer (e.g., 1, -1, -3). To find n, count the number of places that the

decimal point must be moved to give the coefficient, C. If the decimal point must be moved to the left, n is a positive integer; if it must be moved to the right, n is a negative integer.

Example 4.1 Express the two numbers given at the beginning of this chapter (the number of atoms in one gram of carbon and the mass in grams of one carbon atom) in exponential notation.

Solution   For the number:

$$50,150,000,000,000,000,000,000$$

the coefficient is 5.015. To obtain this coefficient, the decimal point must be moved 22 places (count them!) to the *left*. It follows that the exponential number is:

$$5.015 \times 10^{22}$$

Similarly, the coefficient of the number:

$$0.0000000000000000000001994$$

is 1.994. The decimal point must be moved 23 places to the *right* to obtain the coefficient. Therefore, we obtain:

$$1.994 \times 10^{-23}$$

EXERCISES

1. Express the following numbers in exponential notation:
   a. 1000   e. 212.6
   b. one billion   f. 0.189
   c. 0.000001   g. 6.18
   d. 16,220   h. 0.00000007846
2. In each of the following pairs, select the number which is larger.
   a. 3×10^3; 3×10^{-3} d. 6×10^7; 4×10^8
   b. 3×10^3; 10,000   e. 9.6×10^{-3}; 1.5×10^{-2}
   c. 0.0001; 2×10^{-4} f. 21×10^3; 2.1×10^4

4.2 MULTIPLICATION AND DIVISION

One of the principal advantages of exponential notation is that it greatly simplifies the processes of multiplication and division. In applying these processes to exponential numbers, we make use of the fact that *to multiply, we add exponents:*

$$10^1 \times 10^2 = 10^{1+2} = 10^3$$

$$10^6 \times 10^{-4} = 10^{6+(-4)} = 10^2$$

To divide, we subtract exponents:

$$10^3/10^2 = 10^{3-2} = 10^1$$
$$10^{-3}/10^6 = 10^{-3-6} = 10^{-9}$$
$$10^4/10^{-6} = 10^{4-(-6)} = 10^{10}$$

Using these principles, we arrive at the following rules for multiplying or dividing exponential numbers.

To multiply one exponential number by another, first multiply the coefficients together in the usual manner. Then add exponents.

To divide one exponential number by another, divide coefficients in the usual manner and subtract exponents.

Example 4.2 Carry out the indicated operations:
  a. $(5.00 \times 10^4) \times (1.60 \times 10^2)$
  b. $(6.01 \times 10^{-3})/(5.23 \times 10^6)$

Solution

  a. For convenience, we first separate the coefficients from the exponential terms:

$$(5.00 \times 1.60) \times (10^4 \times 10^2)$$

Multiplying coefficients and adding exponents, we obtain: 8.00×10^6. (Here, and throughout this chapter, we use the rules discussed in Chapter 7 to express answers to the correct number of significant figures.)

  b. $(6.01 \times 10^{-3})/(5.23 \times 10^6) = \dfrac{6.01}{5.23} \times \dfrac{10^{-3}}{10^6} = 1.15 \times 10^{-9}$

We frequently find that when exponential numbers are multiplied or divided, our answer is not in standard exponential notation. Consider, for example:

$$(5.0 \times 10^4) \times (6.0 \times 10^3)$$

Carrying out this multiplication in the usual manner, we obtain:

$$(5.0 \times 6.0) \times (10^4 \times 10^3) = 30 \times 10^7$$

Again:

$$(3.60 \times 10^2)/(4.92 \times 10^4) = \frac{3.60}{4.92} \times \frac{10^2}{10^4} = 0.732 \times 10^{-2}$$

The two numbers just obtained, 30×10^7 and 0.732×10^{-2}, are not expressed in standard exponential form, since the coefficients are *not* numbers between 1 and

10. To express these numbers in exponential notation, we follow the procedure described in Example 4.3.

Example 4.3 Express the numbers 30×10^7 and 0.732×10^{-2} in standard exponential notation (i.e., as numbers between 1 and 10, times 10 to the proper power).

Solution In the first case, we write:

$$30 \times 10^7 = (3.0 \times 10^1) \times 10^7$$

Adding exponents, we obtain: 3.0×10^8

What we did here was to divide the coefficient by 10 so as to obtain a number, 3.0, which lies between 1 and 10. To compensate for this, we multiplied the exponential term, 10^7, by 10 to get 10^8.

Following an analogous procedure in the second case:

$$0.732 \times 10^{-2} = (7.32 \times 10^{-1}) \times 10^{-2} = 7.32 \times 10^{-3}$$

Here, we multiplied the coefficient by 10 and divided the exponential by 10, leaving the value of the number unchanged.

From these examples, we draw the general rules:

If the coefficient is less than 1, multiply by the appropriate power of 10 so that it falls between 1 and 10. Divide the exponential by the same power of 10, thereby leaving the expression itself unchanged.

If the coefficient is greater than 10, divide by the appropriate power of 10 so that it falls between 1 and 10. Multiply the exponential by the same power of 10, thereby leaving the expression itself unchanged.

EXERCISES

Carry out the indicated operations, expressing your answers in standard exponential notation.

1. $(6.20 \times 10^4) \times (1.50 \times 10^8)$
2. $(4.3 \times 10^{-3}) \times (9.0 \times 10^4)$
3. $(3.62 \times 10^4) \times (2.91 \times 10^{-7})$
4. $(8.16 \times 10^{-4}) \times (4.78 \times 10^{19})$
5. $(1.39 \times 10^7)/(1.10 \times 10^4)$

6. $(3.48 \times 10^3)/(6.72 \times 10^5)$
7. $(7.2 \times 10^{-3})/(3.6 \times 10^4)$
8. $(2.60 \times 10^4)/(7.70 \times 10^{-12})$
9. $\dfrac{(6.10 \times 10^4) \times (3.18 \times 10^{-4})}{(8.08 \times 10^7) \times (1.62 \times 10^{11})}$

4.3 RAISING TO POWERS AND EXTRACTING ROOTS

To raise an exponential number to a power, we take advantage of the fact that

$$(10^a)^b = 10^{a \times b}$$

To illustrate this rule, consider

$$(10^2)^3 = 10^2 \times 10^2 \times 10^2 = 10^6 = 10^{(2 \times 3)}$$
$$(10^{-2})^4 = 10^{-2} \times 10^{-2} \times 10^{-2} \times 10^{-2} = 10^{-8} = 10^{(-2 \times 4)}$$

To raise an exponential number to a power, we treat the coefficient in the usual manner and raise the exponential term according to the rule just given:

$$(2.0 \times 10^{-3})^2 = (2.0)^2 \times (10^{-3})^2 = 4.0 \times 10^{-6}$$

The same principle can be used to extract a root (square root, cube root, etc.). Here we are dealing with a fractional power:

$$\sqrt[2]{10} = 10^{1/2}; \quad \sqrt[3]{10} = 10^{1/3}. \quad \sqrt[n]{10} = 10^{1/n}$$

but the operation is entirely analogous. Thus:

$$\sqrt[2]{10^6} = (10^6)^{1/2} = 10^{(6 \times 1/2)} = 10^3$$
$$\sqrt[2]{4.0 \times 10^6} = (4.0 \times 10^6)^{1/2} = (4.0)^{1/2} \times (10^6)^{1/2} = 2.0 \times 10^3$$

As before, we operate on the coefficient and exponential separately.

Extracting square roots poses a special problem when the exponent is not an even number, i.e., divisible by 2 to give an integer. Consider, for example:

$$(4.0 \times 10^5)^{1/2}$$

If we follow the procedure outlined previously, we obtain:

$$(4.0)^{1/2} \times (10^5)^{1/2} = 2.0 \times 10^{5/2}$$

Our answer is not in standard exponential form; indeed, $10^{5/2}$ is an extremely awkward expression to work with because it cannot readily be translated into an ordinary number.

In cases of this type, we first transform the number we are working with into a form such that extracting the root of the exponential term will give a whole number. In the preceding example, we write:

$$(4.0 \times 10^5)^{1/2} = (40 \times 10^4)^{1/2}$$

multiplying the coefficient by 10 and dividing the exponential by 10. Now, on extracting the square root, we obtain:

$$(40 \times 10^4)^{1/2} = 40^{1/2} \times (10^4)^{1/2} = 40^{1/2} \times 10^2 = 6.3 \times 10^2$$

(The square root of 40 may be found from tables, by the use of logarithms [Chapter 5], or with the aid of a slide rule [Chapter 6].)

From this example, we draw the general rule that to extract the square root of an exponential number where the exponent is odd (e.g., $-1, 3, 5$), we convert it to an even number by dividing by 10, simultaneously multiplying the coefficient by 10. Thus we have:

$$(1.0 \times 10^{-1})^{1/2} = (10 \times 10^{-2})^{1/2} = 3.2 \times 10^{-1}$$
$$(5.0 \times 10^{3})^{1/2} = (50 \times 10^{2})^{1/2} = 7.1 \times 10^{1}$$

The same principle is used in extracting cube or higher roots: to extract the nth root of an exponential number, we first make sure that the exponent is divisible by n to give an integer. For example, to obtain the cube root of 2.0×10^5, we would rewrite the exponential as either 0.20×10^6 (i.e., multiply the exponent by 10, divide the coefficient by 10) or as 200×10^3 (divide the exponent by 100, multiply the coefficient by 100) before attempting to take the root.

$$(2.0 \times 10^5)^{1/3} = (0.20 \times 10^6)^{1/3} = 0.58 \times 10^2 = 58$$

or:

$$(2.0 \times 10^5)^{1/3} = (200 \times 10^3)^{1/3} = 5.8 \times 10^1 = 58$$

In carrying out a calculation of this sort, you must be sure that the value of the expression in which the exponent has been changed is the same as it was before. If the magnitude of the exponential term is *increased*, the magnitude of the coefficient must be *decreased*, so that their product does not change. If the exponent is positive, increasing its size (i.e., from 10^3 to 10^4) will increase the magnitude of the exponential term, in which case the coefficient must be made smaller. If, on the other hand, the exponent is negative, increasing its size (i.e., from 10^{-3} to 10^{-4}) will make the exponential term smaller; to compensate for this change, the coefficient would have to be made larger.

Example 4.4 Perform the indicated operations.
  a. $(6.2 \times 10^{-4})^2$
  b. $(3.0 \times 10^6)^{1/2}$
  c. $(2.81 \times 10^{-5})^{1/2}$

Solution

  a. $(6.2 \times 10^{-4})^2 = (6.2)^2 \times 10^{-8} = 38 \times 10^{-8} = 3.8 \times 10^{-7}$ (Note that in order to obtain the answer in standard form, we divided the coefficient by 10 and multiplied the exponential term by 10.)
  b. $(3.0 \times 10^6)^{1/2} = (3.0)^{1/2} \times 10^3 = 1.7 \times 10^3$
      (The square root of 3.0 is approximately 1.7.)
  c. Here, we must first convert the expression to get an exponent which gives an integer when divided by 2. One way to do this is to divide the exponential term by 10 and multiply the coefficient by 10.

$$(2.81 \times 10^{-5})^{1/2} = (28.1 \times 10^{-6})^{1/2} = (28.1)^{1/2} \times (10^{-6})^{1/2}$$
$$= 5.30 \times 10^{-3}$$

1. $(2.16 \times 10^{-3})^2 = ?$
2. $(4.9 \times 10^4)^3 = ?$
3. $(6.0 \times 10^{-21})^2 = ?$
4. $(9.0 \times 10^6)^{1/2} = ?$
5. $(8.4 \times 10^5)^{1/2} = ?$

6. $(6.2 \times 10^{-2})^{1/2} = ?$
7. $(1.62 \times 10^{-7})^{1/2} = ?$
8. $(2.14 \times 10^{10})^{1/3} = ?$
9. $(6.0 \times 10^4)^2 \times (3.0 \times 10^{-7})^{1/2} = ?$
10. $(3.0 \times 10^7)^{1/5} = ?$

4.4 ADDITION AND SUBTRACTION

Occasionally, we find it necessary to add or subtract two exponential numbers. These processes are extremely simple if both exponents are the same. To add:

$$2.02 \times 10^7 + 3.16 \times 10^7$$

we factor to obtain: $(2.02 + 3.16) \times 10^7 = 5.18 \times 10^7$

Again,

$$6.1 \times 10^{-5} - 3.0 \times 10^{-5} = (6.1 - 3.0) \times 10^{-5} = 3.1 \times 10^{-5}$$

If the exponents are different, the numbers must be operated upon to make the exponents the same. This procedure is illustrated in Example 4.5.

Example 4.5 Carry out the indicated operations.
    a. $6.04 \times 10^3 + 2.6 \times 10^2$
    b. $9.82 \times 10^{-6} - 8.2 \times 10^{-5}$

Solution

    a. We cannot perform the addition directly, any more than we can add six oranges to two apples. In order to carry out a meaningful addition, we must make the exponents of the two terms the same. One way to do this is to operate on the second number, expressing it as a coefficient times 10^3.

$$2.6 \times 10^2 = 0.26 \times 10^3$$

Therefore:

$$6.04 \times 10^3 + 2.6 \times 10^2 = 6.04 \times 10^3 + 0.26 \times 10^3 = 6.30 \times 10^3$$

    b. Proceeding as in part (a):

$$9.82 \times 10^{-4} - 8.2 \times 10^{-5} = 9.82 \times 10^{-4} - 0.82 \times 10^{-4}$$

$$= 9.00 \times 10^{-4}$$

Alternatively, we could have operated on the first number rather than the second:

$$9.82 \times 10^{-4} = 98.2 \times 10^{-5}$$

Hence:

$$9.82 \times 10^{-4} - 8.2 \times 10^{-5} = 98.2 \times 10^{-5} - 8.2 \times 10^{-5}$$
$$= 90.0 \times 10^{-5} = 9.00 \times 10^{-4}$$

EXERCISES

Carry out the indicated additions and subtractions.
1. $3.02 \times 10^4 + 1.69 \times 10^4$
2. $4.18 \times 10^{-2} + 1.29 \times 10^{-2}$
3. $6.10 \times 10^4 + 1.0 \times 10^3$
4. $5.9 \times 10^{-5} + 1.86 \times 10^{-4}$
5. $8.17 \times 10^5 - 1.20 \times 10^4$
6. $6.49 \times 10^{-10} - 1.23 \times 10^{-11}$
7. $9.68 \times 10^4 + 7.01 \times 10^2$
8. $6.02 \times 10^{23} - 1.0 \times 10^2$

4.5 GENERAL RULES

Throughout this chapter, we have emphasized the rules for operating on exponential numbers, which appear so frequently in general chemistry. In all such numbers, the base of the exponent is, of course, 10. However the rules that we have discussed apply to operations involving any exponent, regardless of base. Thus, not only does

$$10^2 \times 10^3 = 10^5; \quad 10^2/10^3 = 10^{-1}$$

but also:

$$6^2 \times 6^3 = 6^5; \quad 7^2/7^3 = 7^{-1}$$

The general rules and conventions for working with exponents to any base are summarized in Table 4.2.

Table 4.2 General Rules for Use of Exponents

General	Example (base 10)	Example (base 2)	Exercise
1. $Y^n = Y$ multiplied by itself n times $(n \neq 0)$	$10^3 = 10 \times 10 \times 10$	$2^4 = 2 \times 2 \times 2 \times 2$	$3^5 = ?$
2. $Y^0 = 1$	$10^0 = 1$	$2^0 = 1$	$7. 2^0 = ?$
3. $Y^{-n} = 1/Y^n$	$10^{-3} = 1/10^3$	$2^{-4} = 1/2^4$	$3^{-2} = ?$
4. $Y^{1/n} = \sqrt[n]{Y}$	$10^{1/2} = \sqrt[2]{10}$	$2^{1/3} = \sqrt[3]{2}$	$6. 4^{1/5} = ?$
5. $Y^n \times Y^m = Y^{(n+m)}$	$10^2 \times 10^3 = 10^5$	$2^2 \times 2^4 = 2^6$	$3^2 \times 3^{-4} = ?$
6. $Y^n/Y^m = Y^{(n-m)}$	$10^2/10^3 = 10^{-1}$	$2^2/2^4 = 2^{-2}$	$8^4/8^1 = ?$
7. $(Y^n)^m = Y^{nm}$	$(10^3)^2 = 10^6$	$(2^2)^3 = 2^6$	$(7^3)^{-2} = ?$
8. $(Y^n)^{1/m} = Y^{n/m}$	$(10^4)^{1/2} = 10^{4/2}$	$(2^4)^{1/3} = 2^{4/3}$	$(5^3)^{1/2} = ?$

Frequently in general chemistry we find units as well as numbers raised to powers. For example, we may record the volume of a flask as 125 cm^3. In another case, the density of mercury may be given as 13.6 g/cm^3. Negative as well as positive exponents may be used; we might, for example, quote the density of mercury as 13.6 g cm^{-3}. Again, the gas constant R may be written as:

$$8.21 \times 10^{-2} \text{ lit atm/mole } °K$$

or alternatively as:

$$8.21 \times 10^{-2} \text{ lit atm mole}^{-1} (°K)^{-1}$$

Units expressed in this manner can be operated on like any other exponents (Example 4.6).

Example 4.6 The rate of a certain reaction is given by the expression:

$$\text{rate} = k \, (\text{conc. } A)^2$$

Rate has the units mole^1 liter^{-1} sec^{-1}, concentration the units mole^1 liter^{-1}. What are the units of the rate constant, k?

Solution Solving for k:

$$k = \text{rate}/(\text{conc. } A)^2$$

Substituting units:

$$\frac{\text{mole}^1 \text{ liter}^{-1} \text{ sec}^{-1}}{(\text{mole}^1 \text{ liter}^{-1})^2} = \frac{\text{mole}^1 \text{ liter}^{-1} \text{ sec}^{-1}}{\text{mole}^2 \text{ liter}^{-2}} = \text{mole}^{-1} \text{ liter}^1 \text{ sec}^{-1}$$

PROBLEMS

These problems are designed to illustrate the use of exponential numbers in general chemistry. Don't panic if you are unfamiliar with some of the vocabulary that is used. A knowledge of exponents combined with a little common sense will enable you to work all of the problems. You may even learn some chemistry in the process!

The answers to the problems are given in Appendix 3, in standard exponential notation.

4.1 A helium atom has the following properties:
    mass = 0.000000000000000000000000665 g
    radius = 0.0000000093 cm
    average velocity at 25°C = 136,000 cm/sec
    Express these quantities in exponential notation.

4.2 The solubility of lead sulfate in water at 25°C is 1.0×10^{-4} mole/liter. One mole of lead sulfate weighs 303 grams. Express the solubility in grams/liter, giving your answer first in standard exponential notation and then as an ordinary number.

4.3 When one gram of carbon burns to form carbon dioxide, 7.86×10^3 calories of heat is evolved. Calculate the amount of heat evolved when one mole (12.0 g) of carbon burns. This quantity is called the molar heat of combustion.

4.4 One mole (6.02×10^{23} molecules) of water weighs 18.0 g. Calculate the mass in grams of a water molecule.

4.5 The half life, $t_{1/2}$, of a radioactive isotope is the time required for one half of a sample to decay. It is related to the rate constant for radioactive decay, k, by the equation:

$$t_{1/2} = 0.693/k$$

For the common isotope of radium, $k = 4.33 \times 10^{-4}$/yr. What is the half life of this isotope in:
  a. Years
  b. Days
  c. Minutes

4.6 The concentration of sulfur dioxide in polluted air is 2.0×10^{-7} grams/liter.
  a. How many grams of sulfur dioxide are there in 200 ml of this air (1 liter = 10^3 ml)?
  b. How many liters of air must be taken to yield one gram of sulfur dioxide?

4.7 The solubility of helium in water at 25°C and one atm. is 3.8×10^{-5} moles He/cc water. One mole of helium weighs 4.0 g and has a volume of 2.5×10^4 cc at 25°C. Express the solubility of helium in:
  a. g He/cc water
  b. cc He/cc water

4.8 In a solution formed by dissolving hypochlorous acid, HClO, in water, the following relation exists between the concentrations of H^+ and HClO:

$$(\text{conc. } H^+)^2 = (3.2 \times 10^{-8})(\text{conc. HClO})$$

Complete the following table:

(conc. HClO)	(conc. H^+)	(conc. H^+)/(conc. HClO)
1.0	———	———
0.10	———	———
0.010	———	———

The ratio in the last column represents the fraction of HClO which ionizes.

4.9 According to the Bohr theory of the hydrogen atom, the orbital radius is given by:

$$r = \frac{n^2 h^2}{4\pi^2 m e^2}$$

where r is the radius in cm, $h = 6.6 \times 10^{-27}$, $e = 4.8 \times 10^{-10}$, $m = 9.1 \times 10^{-28}$, $\pi = 3.14$, and n is the so-called quantum number which can take on any positive, integral value. Calculate r when $n = 1$; $n = 2$; $n = 3$.

4.10 In a solution prepared by saturating pure water with $Ag_2 CrO_4$, the concentration of CrO_4^{2-} is 6.0×10^{-5}. The concentration of Ag^+ is exactly twice that of CrO_4^{2-}. Calculate the solubility product, K_{sp}, of $Ag_2 CrO_4$, defined by the equation:

$$K_{sp} = (\text{conc. } Ag^+)^2 \times (\text{conc. } CrO_4^{2-})$$

4.11 The metal calcium crystallizes in a cubic structure in which the face diagonal, d, of the cube is 7.88×10^{-8} cm. Calculate the length, l, of a side of the cube, using the relation $2l^2 = d^2$.

4.12 For the three compounds AgBr, PbF_2, and $Fe(OH)_3$, the solubility s is given by the expression:

AgBr: $s^2 = 1.0 \times 10^{-13}$

PbF_2: $4s^3 = 4.0 \times 10^{-8}$

$Fe(OH)_3$: $27s^4 = 5.0 \times 10^{-38}$

Calculate the solubility of each compound.

4.13 When ammonia is added to silver chloride, AgCl, some of the Ag^+ ions are converted to the $Ag(NH_3)_2^+$ complex. At equilibrium, the following equation holds:

$$\frac{(\text{conc. } Ag^+) \times (\text{conc. } NH_3)^2}{(\text{conc. } Ag(NH_3)_2^+)} = 4 \times 10^{-8}$$

What must the concentration of NH_3 be to make:
   a. conc. $Ag(NH_3)_2^+$ = conc. Ag^+
   b. conc. $Ag(NH_3)_2^+$ = 10^2 (conc. Ag^+)

4.14 A sample of water vapor occupying 2.06×10^4 cc is in equilibrium with 1.2×10^3 cc of liquid water. What is the total volume of the system?

4.15 Dalton's Law tells us that the total pressure of a mixture of hydrogen and helium is the sum of the partial pressures of the two gases. If the total pressure of the mixture is 1.224×10^3 torr and the partial pressure of helium is 9.80×10^2 torr, what is the partial pressure of hydrogen?

4.16 For a certain reaction between two molecules A and B, the rate equation is:

$$\text{rate} = k \,(\text{conc. } A)(\text{conc. } B)^2$$

Concentration has the units mole liter^{-1}; the rate is expressed as mole liter^{-1} min^{-1}. What are the units of the rate constant, k?

CHAPTER 5

LOGARITHMS

It was pointed out in Chapter 4 that the processes of multiplication, division, raising to a power, and extracting a root are simplified by expressing the numbers involved in exponential notation. However, even when this is done, a considerable amount of arithmetic may still be necessary. For example, suppose we wish to multiply 6.02×10^{23} by 1.99×10^{-24}. Combining the exponential terms, we arrive at the expression:

$$(6.02 \times 1.99) \times 10^{-1}$$

but we still have to carry out the tedious operation of multiplying 6.02 by 1.99. We could achieve a further simplification if we could express the numbers 6.02 and 1.99 as powers of ten. The time consuming operation of multiplication could then be completely replaced by the more rapid process of addition.

Clearly, what we need is a table which will enable us to express any number as a power of ten. Such a table has been known to mathematicians for more than three centuries. It is known as a table of **common logarithms** and is reproduced in Appendix 2. A common logarithm is simply a power of 10; specifically, **it is the power to which 10 must be raised to give a particular number.** In other words:

$$\text{if } x = 10^{y}, \text{ then } y = \log x \qquad (5.1)^*$$

To understand what these statements mean, let us examine Table 5.1. Notice that numbers such as 0.01, 1, and 100, which can be expressed as integral powers of ten ($0.01 = 10^{-2}$, $1 = 10^{0}$, $100 = 10^{2}$) have integral logarithms (-2, 0, $+2$). In contrast, the numbers 2–9 in the center column of Table 5.1, none of which is an inte-

---

*Numbers other than 10 can be used as a base for logarithms (see Section 5.4). As the phrase "common logarithm" implies, 10 is the base most commonly used. When we use the word "logarithm" or its abbreviation "log" without any qualifier, we refer to base 10 logarithms. In cases where there might be some ambiguity about the base, we shall write "$\log_{10}x$" rather than simply "$\log x$" to represent the common logarithm.

Table 5.1   Logarithms of a Few Numbers Between 0.001 and 1000

x	$\log x$	x	$\log x$	x	$\log x$
0.001	-3	2	0.3010	10	1
0.01	-2	3	0.4771	100	2
0.1	-1	4	0.6021	1000	3
1	0	5	0.6990		
		6	0.7782		
		7	0.8451		
		8	0.9031		
		9	0.9542		

gral power of 10, do not have integral logarithms. Concentrating upon the number 3, we see that its logarithm is, to four decimal places, 0.4771. This means that:

$$10^{0.4771} = 3.000$$

Notice that since the number 3 is intermediate between the numbers 1 and 10, its logarithm (0.4771) is intermediate between those of 1, whose log is 0, and 10, whose log is 1. Indeed, since 3 is a little less than the square root of 10 ($10^{1/2} = 3.162$), it is not surprising that its logarithm should be a little less than $1/2$.

In Figure 5.1, we have plotted a portion of the data given in Table 5.1. Notice

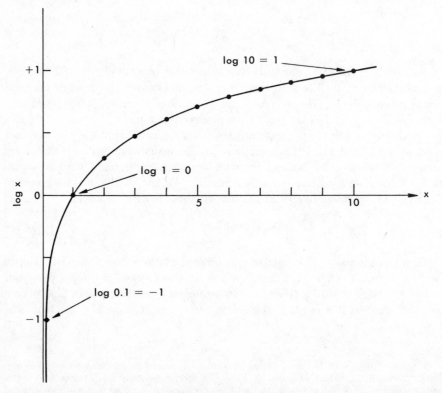

FIGURE 5.1 Numbers greater than 1 have positive logarithms. Numbers less than 1 have negative logarithms.

that the curve crosses the horizontal axis (log x = 0) at x = 1. This means that numbers greater than 1 have positive logarithms, while numbers less than 1 have negative logarithms.

$$\text{if } x > 1, \text{ then } \log x > 0$$
$$\text{if } 0 < x < 1, \text{ then } \log x < 0$$

(5.2)

As the number approaches zero, its logarithm approaches negative infinity. Numbers less than zero, such as -1 or -6.82, cannot be assigned logarithms. It is impossible to obtain a negative number by raising 10 to any power whatsoever.

5.1 FINDING THE LOGARITHM OF A NUMBER

The four-place table of logarithms given in Appendix 2 allows us to determine directly the logarithm of any three-digit number between 1 and 10. To illustrate, suppose we wish to find the logarithm of 3.84. We follow down the column at the far left of the table until we come to 3.8 and then move across to the column headed "4," reading off the logarithm of 3.84 as 0.5843. Similarly, we would find the logarithm of 3.85 to be 0.5855, that of 3.86 to be 0.5866, and so on.

It is also possible to estimate quite accurately from the table the logarithms of four-digit numbers. Suppose, for example, we want to know the logarithm of 3.845. We know that the logarithms of 3.84 and 3.85 are 0.5843 and 0.5855, respectively. Since the number 3.845 is half way between 3.84 and 3.85, its logarithm should be about half way between 0.5843 and 0.5855, i.e., 0.5849. Expressing our reasoning in mathematical language:

$$\log 3.845 = \log 3.84 + 0.5 (\log 3.85 - \log 3.84)$$
$$= 0.5843 + 0.5(0.5855 - 0.5843)$$
$$= 0.5843 + 0.5(0.0012) = 0.5849$$

Example 5.1 Find the logarithms of all the four-digit numbers between 2.000 and 2.010 (i.e., 2.001, 2.002, 2.003, $\cdots$ 2.009).

Solution We note that the difference between the logs of 2.010 and 2.000 is:

$$0.3032 - 0.3010 = 0.0022$$

Therefore:

$$\log 2.001 = 0.3010 + 0.1(0.0022) = 0.3012$$
$$\log 2.002 = 0.3010 + 0.2(0.0022) = 0.3014$$

Proceeding similarly with the other four-digit numbers, we obtain:

2.001	0.3012	2.004	0.3019	2.007	0.3025
2.002	0.3014	2.005	0.3021	2.008	0.3028
2.003	0.3017	2.006	0.3023	2.009	0.3030

A table of logarithms can also be used to find logs of numbers less than 1 or greater than 10. Consider, for example, the number 384. Writing 384 in exponential notation, we obtain 3.84×10^2. But, since a logarithm is an exponent, the log of this product must be the sum of the logs of 3.84 and 10^2. That is:

$$\log (3.84 \times 10^2) = \log 3.84 + \log 10^2$$

Recalling that the log of 3.84 is 0.5843 and that the log of 10^2 must be 2, we have:

$$\log(3.84 \times 10^2) = 0.5843 + 2 = 2.5843$$

Similarly, to find the logarithm of 0.00384:

$$\log(0.00384) = \log(3.84 \times 10^{-3}) = 0.5843 - 3 = -2.4157$$

In the general case:

$$\log(C \times 10^n) = n + \log C \tag{5.3}$$

Example 5.2   Find the logarithms of 6.02×10^{23} and 0.602.

Solution   From Appendix 2, we read directly the log of 6.02 to be 0.7796. Hence:

$$\log(6.02 \times 10^{23}) = 23 + \log 6.02$$

$$= 23 + 0.7796 = 23.7796$$

$$\log 0.602 = \log(6.02 \times 10^{-1})$$

$$= -1 + \log 6.02 = -1 + 0.7796 = -0.2204$$

EXERCISES

Find the logarithms of the following numbers:

1. 8.16	3. 1.652	5. 56.52	7. 0.004918
2. 1.03	4. 5.841	6. 4.18×10^9	8. 3.87×10^{-12}

5.2 FINDING THE NUMBER CORRESPONDING TO A GIVEN LOGARITHM

The operation of finding an antilogarithm (number corresponding to a given logarithm) is simply the inverse of finding the logarithm of a number. Experience has shown, however, that students often have difficulty with this operation.

To start with a simple example, let us find the number whose logarithm is 0.4997. To do this, we locate 0.4997 in the body of the table. We note that it falls in the horizontal column labeled "3.1" under the vertical column labeled "6." We deduce that the antilogarithm of 0.4997 is 3.16 (i.e., 3.16 is the number whose logarithm is 0.4997). By the same procedure, we could deduce that the antilogarithm of 0.5011 is 3.17.

Now let us consider the slightly more complicated (and much more common) case, in which the logarithm we are working with does not appear directly in the table. Suppose, for example, we wish to find the antilogarithm of 0.5000. Although we cannot locate this logarithm directly in the table, we can bracket it between 0.4997, the logarithm of 3.16, and 0.5011, the logarithm of 3.17. We deduce that since 0.5000 lies between 0.4997 and 0.5011, its antilogarithm must lie between 3.16 and 3.17. Specifically, 0.5000 is $^3/_{14}$ or about 0.2 of the way between 0.4997 and 0.5011; we estimate that its antilogarithm is 0.2 of the way between 3.16 and 3.17. Since 3.162 is the number which lies 0.2 of the way from 3.16 to 3.17, it must be the antilogarithm of 0.5000.

Example 5.3 Find the number whose logarithm is 0.8641.

Solution Scanning the table, we find that:

$$0.8639 = \log 7.31; \quad 0.8645 = \log 7.32$$

Since 0.8641 lies $^2/_6$ or approximately 0.3 of the way between 0.8639 and 0.8645, it follows that the number must be 0.3 of the way between 7.31 and 7.32. In other words, the antilogarithm of 0.8641 must be approximately 7.313.

The process which we have just described can be applied to logarithms such as 0.5000 or 0.8641, which fall between 0 and 1. Frequently, we need to find numbers corresponding to logarithms which are greater than 1 (e.g., 1.5000, 25.8641) or less than zero (-0.5000, -2.1359). The general principle which we shall use in all such cases is to **rewrite the logarithm so that it is in the form of a positive decimal fraction (mantissa) plus or minus a whole number (characteristic).**

To illustrate this procedure, consider the problem of finding the number whose logarithm is 25.8641. We cannot, of course, find this logarithm directly in the table, which is limited to logarithms having values between 0.0000 and 1.0000. Consequently, we rewrite 25.8641 as 0.8641 + 25. Now, we look up the mantissa, 0.8641, in the table, and find that its antilogarithm is 7.313 (see Example 5.3). The antilogarithm of the characteristic, 25, is 10^{25}. It follows that the number we are looking for is 7.313×10^{25}. Mathematically:

$$\log(7.313 \times 10^{25}) = \log 7.313 + \log 10^{25}$$

$$= 0.8641 + 25 = 25.8641$$

You will note that what we have done is to **use the mantissa to determine the coefficient of the exponential number; the characteristic simply tells us the exponent.**

If we are dealing with a logarithm which is a negative number, we proceed in an entirely analogous manner. Let us suppose that we wish to find the number whose logarithm is -3.6990. We must first rewrite -3.6990 to get it in the form of a positive decimal fraction between 0.0000 and 1.0000, minus a whole number. Referring to Figure 5.2, we see that -3.6990 lies 0.3010 units above -4. In other words:

$$-3.6990 = 0.3010 - 4$$

Having passed this hurdle, we proceed as before, noting that the antilogarithm of 0.3010 is 2.000 and that of -4 is 10^{-4}. Consequently, the number we are looking for must be 2.000×10^{-4}.

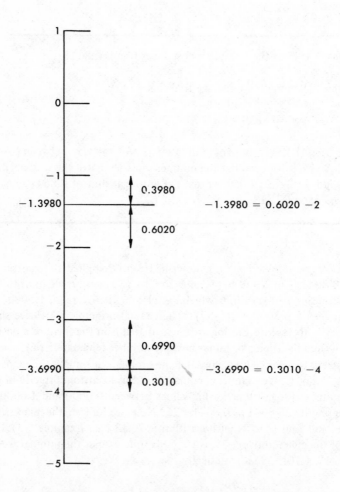

FIGURE 5.2 Writing negative logarithms in standard form.

Example 5.4 Find the numbers whose logarithms are:
   a. 6.4771   b. -1.3979   c. -0.4097

Solution

   a. Antilog 6.4771 = antilog (0.4771 + 6).
      Looking up 0.4771 in the table, we find its antilog to be 3.000;
      the antilog of 6 is 10^6. Consequently:

$$\text{antilog}(0.4771 + 6) = 3.000 \times 10^6$$

   b. Rewriting -1.3979 as a decimal fraction minus a whole number,
      we have:

$$-1.3979 = 0.6021 - 2$$

      The antilog of 0.6021 is 4.000; that of -2 is 10^{-2}. Hence, the
      desired number is 4.000×10^{-2}.
   c. Here, as in (b), we first put the number in standard form:

$$-0.4097 = 0.5903 - 1$$

      In this case, we cannot locate 0.5903 directly in the table. How-
      ever, we see that it falls between 0.5899, the antilog of 3.89, and
      0.5911, the antilog of 3.90. Since 0.5903 is 4/12 or about 0.3
      of the way from 0.5899 to 0.5911, its antilog must be about 0.3
      of the way from 3.89 to 3.90, i.e., 3.893. Hence:

$$\text{antilog}(0.5903 - 1) = 3.893 \times 10^{-1}$$

EXERCISES

1. Find the numbers whose logarithms are:
   a. 0.8831   b. 0.9367   c. 1.7435   d. 1.6165   e. 7.6221
2. Rewrite the following logarithms as decimal fractions minus whole numbers:
   a. -0.6576 b. -2.4023 c. -2.6195 d. -12.4000
3. Find the antilogarithms of the logs given in (2).

5.3 OPERATIONS INVOLVING LOGARITHMS

Since logarithms are exponents, the rules that we derived in Chapter 4 for performing mathematical operations with exponents can be extended to logarithms.
The results are summarized in Table 5.2.

Table 5.2   Mathematical Operations Involving Exponents and Logarithms

	Exponents	Logarithms
Multiplication	$10^a \times 10^b = 10^{(a+b)}$	$\log (xy) = \log x + \log y$
Division	$10^a/10^b = 10^{(a-b)}$	$\log x/y = \log x - \log y$
Raising to a power	$(10^a)^n = 10^{an}$	$\log x^n = n \log x$
Extracting a root	$(10^a)^{1/n} = 10^{a/n}$	$\log x^{1/n} = \dfrac{1}{n} \log x$

Example 5.5 Using Table 5.2 and a table of logarithms, calculate:

a. $\log (2.061 \times 4.190)$

b. $\log \dfrac{3.160 \times 10^4}{2.082 \times 10^5}$

c. $\log (6.023)^3$

Solution

a. $\log 2.061 = 0.3141; \quad \log 4.190 = 0.6222$
    $\log (2.061 \times 4.190) = 0.3141 + 0.6222 = 0.9363$

b. $\log 3.160 \times 10^4 = 4.4997; \quad \log 2.082 \times 10^5 = 5.3185$
    $\log \dfrac{3.160 \times 10^4}{2.082 \times 10^5} = 4.4997 - 5.3185 = 0.1812 - 1 \ (\text{or} -0.8188)$

c. $\log (6.023)^3 = 3 \log (6.023) = 3(0.7798) = 2.3394$

The rules listed in Table 5.2 suggest a way of simplifying certain mathematical processes. They enable us to substitute the simpler processes of addition and subtraction for the tedious operations of multiplication and division. Suppose, for example, we wish to multiply 2.061 by 4.190. Following the procedure shown in Example 5.5a, we have:

$$\log(2.061 \times 4.190) = 0.3141 + 0.6222 = 0.9363$$

Using a table of logarithms, we find the antilog of 0.9363 to be 8.636. Consequently, we deduce that:

$$2.061 \times 4.190 = 8.636$$

Similarly, to divide 3.160×10^4 by 2.082×10^5, we have:

$$\log \frac{3.160 \times 10^4}{2.082 \times 10^5} = 4.4997 - 5.3185 = 0.1812 - 1$$

$$\text{antilog } 0.1812 - 1 = 1.518 \times 10^{-1}$$

(Note that in this case, it was more convenient to leave the logarithm in the form $0.1812 - 1$, rather than writing it as -0.8188.)

In practice, most of the multiplications and divisions that we carry out in general chemistry can be accomplished with sufficient accuracy on a slide rule (Chapter 6).

Since a slide rule is generally more convenient to use than a table of logarithms, we shall not dwell further on the use of logarithms for multiplication and division. A slide rule can also be used for squaring and cubing numbers and for extracting square or cube roots. However, raising a number to a large power (e.g., 6, or 10) or to a very small fractional power (1/4, 1/5) can be extremely tedious. Operations such as these, which arise in a variety of practical calculations, are perhaps most conveniently carried out using logarithms (Example 5.6).

Example 5.6 The total value of a sum of money, S, invested at interest compounded annually is:

$$S(1 + i/100)^n$$

where i is the per cent interest and n is the number of years. If one dollar is invested in a bank that pays 5% per year, compounded annually, how much does it amount to after 20 years?

Solution Here, $S = 1.00$, $i/100 = 0.05$, and $n = 20$. Letting d stand for the value of the deposit after 20 years:

$$d = 1.00(1.05)^{20}$$

Taking logarithms:

$$\log d = \log 1.00 + 20 \log 1.05$$

$$= 0.0000 + 20(0.0212) = 0.4240$$

The antilog of 0.4240 is approximately 2.65. In other words, after 20 years at 5% interest compounded annually, $1.00 has increased to $2.65.

EXERCISES

Making use of a table of logarithms, find:

1. $\log(6.160 \times 10^3)(1.680 \times 10^{-2})$

2. $\log \dfrac{4.983 \times 10^4}{3.172 \times 10^2}$

3. $\log \dfrac{1.684 \times 10^5}{6.480 \times 10^7}$

4. $\log(9.080 \times 10^2)^3$

5. $\log(6.162)^{1/4}$

6. $\log \dfrac{(6.161 \times 10^8)(3.812)^2}{(1.976 \times 10^{-10})(6.180)^{1/2}}$

7. $(7.074)^{19}$

8. $\dfrac{(3.265 \times 10^4)^{1/6}}{(6.010 \times 10^{-2})^5}$

9. The number of years required for a deposit at compound interest of 5% to double in value (refer to the formula given in Example 5.6).

5.4 NATURAL LOGARITHMS

To this point, we have been discussing "common logarithms," i.e., logarithms to the base 10. For calculation purposes, common logarithms are the simplest to work with, since our number system is based on multiples of 10. However, certain of the equations which we use in general chemistry are expressed most simply in terms of a different type of logarithm, taken to the base e,* where:

$$e = 2.718 \cdots$$

Logarithms to the base e are referred to as natural logarithms; to distinguish them from common logarithms, the abbreviation $\ln$ is used:

$$\log_e x \equiv \ln x; \qquad \log_{10} x \equiv \log x$$

Tables of natural logarithms are available but, in practice, are seldom used. If we need the natural logarithm of a number, we first look up the base 10 logarithm and then make use of the equation:[†]

$$\ln x = 2.303 \log x \tag{5.4}$$

Example 5.7 Find the natural logarithm of:
   a. 2.280×10^3
   b. 9.831×10^{-2}

Solution In both cases, we first find the base 10 log and then multiply by 2.303.
   a. $\log(2.280 \times 10^3) = 3.3579$;
      $\ln(2.280 \times 10^3) = (2.303)(3.3579) = 7.733$
   b. $\log(9.831 \times 10^{-2}) = 0.9926 - 2 = -1.0074$;
      $\ln(9.831 \times 10^{-2}) = (2.303)(-1.0074) = -2.320$

---

*e can be expressed as an infinite series:

$$e = 1 + \frac{1}{1} + \frac{1}{1 \times 2} + \frac{1}{1 \times 2 \times 3} + \cdots$$

Natural logarithms can be evaluated by summing any one of several different series; one which is particularly convenient because it converges relatively rapidly is:

$$\ln x = 2 \left[\left(\frac{x-1}{x+1} \right) + \frac{1}{3} \left(\frac{x-1}{x+1} \right)^3 + \frac{1}{5} \left(\frac{x-1}{x+1} \right)^5 + \cdots \right]$$

From such series it was relatively easy for mathematicians to compile tables of natural logarithms; in contrast, the direct evaluation of base 10 logs is extremely tedious.

[†]To derive this equation, let $x = 10^y$. We first take the common and then the natural logarithm of both sides of this equation

$$\log x = y \tag{1}$$

$$\ln x = y \ln 10 \tag{2}$$

Substituting for y in (2): $\ln x = (\log x) \ln 10$
But one can show from the series in the footnote above that $\ln 10 = 2.303$
Hence:

$$\ln x = 2.303 \log x$$

Occasionally, we need to evaluate expressions in which the base of natural logarithms, e, is raised to a power (e.g., $e^{0.2500}$). This can readily be accomplished with the aid of Equation 5.4 and a table of common logarithms. The procedure is indicated in Example 5.8.

Example 5.8 Find the numerical value of $e^{0.2500}$

Solution If we let $x = e^{0.2500}$

$$\ln x = 0.2500$$

$$2.303 \log x = 0.2500; \quad \log x = 0.2500/2.303 = 0.1086$$

Taking antilogarithms: $x = 1.284$

EXERCISES

Evaluate:
1. $\ln 6.023$.   2. $\ln (2.02 \times 10^3)$.   3. $\ln (6.18 \times 10^{-5})$.
4. The base 10 log of the number whose natural logarithm is one.
5. The base 10 log of the number whose natural logarithm is 6.190.
6. $e^{22.96}$ 7. $e^{-1.000}$
8. e and $\ln 2$ to three decimal places, using the formulas at the bottom of p. 68.

5.5 APPLICATIONS IN CHEMISTRY

Many of the equations that we work with in general chemistry involve logarithms. Some of the more important of these are listed in Table 5.3. Note that in

Table 5.3 Some Equations of General Chemistry Involving Logarithmic Terms

Equation	Meaning of Symbols
1. $\Delta G^0 = -2.303\ RT \log K$	ΔG^0 = std. free energy change (calories) R = gas constant = 1.987 cal/mole $^\circ$K T = temperature in degrees Kelvin K = equilibrium constant
2. $\log \dfrac{X_0}{X} = \dfrac{kt}{2.303}$ (1st order rate law)	X_0 = original concentration of reactant X = concentration at time t k = 1st order rate constant
3. $\log \dfrac{P_2}{P_1} = \dfrac{\Delta H(T_2 - T_1)}{2.303\ RT_2 T_1}$	P_2, P_1 = vapor pressure of liquid at absolute temperatures T_2 and T_1 ΔH = heat of vaporization of liquid in cal/mole
4. $\log \dfrac{K_2}{K_1} = \dfrac{\Delta H(T_2 - T_1)}{2.303\ RT_2 T_1}$	K_2, K_1 = equilibrium constants at temperatures T_2 and T_1 ΔH = enthalpy change in reaction in calories
5. $\log \dfrac{k_2}{k_1} = \dfrac{\Delta E_a(T_2 - T_1)}{2.303\ RT_2 T_1}$	k_2, k_1 = rate constants at temperatures T_2 and T_1 ΔE_a = activation energy

each of these equations, the number 2.303 appears, reflecting the fact that the relationship would be expressed most simply in terms of natural logarithms. Several of the problems at the end of this chapter (5.6–5.10) illustrate the use of these equations to solve problems in general chemistry.

Logarithms are extensively used in the area of acid-base chemistry. Here we make use of the term pH, defined as:

$$pH = -\log(\text{conc. } H^+) \qquad (5.5)$$

You will have occasion frequently to calculate the pH of solutions of known concentration (Example 5.9). The reverse calculation, which most students find the more difficult of the two, is illustrated in Example 5.10.

Example 5.9 Calculate the pH of solutions in which the concentration of H^+ is

    a. 1×10^{-4} b. 2×10^{-3} c. 5.5×10^{-8}

Solution In each case, we substitute directly in Equation 5.5.

    a. $pH = -\log(1 \times 10^{-4}) = -(-4.0) = 4.0$
    b. $pH = -\log(2 \times 10^{-3}) = -(\log 2 + \log 10^{-3}) = -(0.3 - 3.0) = 2.7$
    c. $pH = -\log(5.5 \times 10^{-8}) = -(\log 5.5 + \log 10^{-8})$
         $= -(0.74 - 8.00) = 7.16*$

Example 5.10 Calculate conc. H^+ in solutions of the following pH:

    a. 6.0   b. 5.3   c. 4.13

Solution Again, we substitute directly in Equation 5.5.

    a. $\log(\text{conc. } H^+) = -6.0$
        Since the antilog of -6.0 is 1×10^{-6}, conc. $H^+ = 1 \times 10^{-6}$.
    b. $\log(\text{conc. } H^+) = -5.3$
        Here, we proceed as usual with negative logarithms, converting -5.3 to a decimal fraction minus a whole number: $-5.3 = 0.7 - 6$

$$\log(\text{conc. } H^+) = 0.7 - 6$$

        From the log table, we see that the antilog of 0.7 is about 5; that of -6 is 10^{-6}.

$$\text{conc. } H^+ = 5 \times 10^{-6}$$

    c. $\log(\text{conc. } H^+) = -4.13 = 0.87 - 5$

$$\text{conc. } H^+ = 7.4 \times 10^{-5}$$

---

*See Chapter 7 for the rules governing the use of significant figures with logarithms and antilogarithms.

The "p ⋯ " notation is used occasionally for terms other than the concentration of H^+ ions. Thus we may refer to "pOH" or "pK," defined as:

$$pOH = -\log(\text{conc. } OH^-); \quad OH^- = \text{hydroxide ion}$$

$$pK = -\log K; \quad K = \text{equilibrium constant}$$

PROBLEMS

5.1 Calculate the pH of solutions which have the following concentrations of H^+:
 a. 1.0×10^{-6} c. 3.0×10^{-9}
 b. 2.61×10^{-2} d. 6.0

5.2 Calculate the concentration of H^+ in solutions of the following pH:
 a. 4.0 c. 3.14
 b. 12.60 d. -1.0

5.3 In a 0.10 M solution of acetic acid, the concentration of H^+ is given by the expression:

$$(\text{conc. } H^+)^2 = 1.80 \times 10^{-6}$$

What is the pH of this solution?

5.4 In a solution saturated with hydrogen sulfide, the following relation holds:

$$(\text{conc. } H^+)^2 \times (\text{conc. } S^{2-}) = 1 \times 10^{-23}$$

What is the concentration of S^{2-} in a solution of this type which has a pH of 4.0?

5.5 In any water solution at $25°C$: $(\text{conc. } H^+) \times (\text{conc. } OH^-) = 1.0 \times 10^{-14}$
 a. What is the pH of a solution in which conc. $OH^- = 2.0 \times 10^{-5}$?
 b. What is conc. OH^- in a solution of pH 8.3?
 c. Show that pH + pOH = 14.0.

5.6 Using Equation 1, Table 5.3, calculate:
 a. ΔG^0 at $300°K$ for a reaction for which $K = 0.020$.
 b. The equilibrium constant for a reaction at $400°K$ where $\Delta G^0 = +4.2$ kcal.

5.7 Using Equation 2, Table 5.3, calculate:
 a. The time required for the concentration of reactant to drop from 1.0 mole/liter to 0.10 mole/liter if $k = 0.045 \text{ min}^{-1}$.
 b. The concentration of reactant remaining after 50 minutes, starting with a concentration of 2.00 mole/liter and taking $k = 2.0 \times 10^{-3}$/minute.

5.8 The vapor pressure of water as a function of temperature is given by Equation 3, Table 5.3, where ΔH_{vap} = 10,200 calories. The vapor pressure of water at $100°C$ is 760 mm Hg.

a. What is the vapor pressure of water at $25°C$ ($298°K$)?

b. At what temperature will the vapor pressure of water be 1.20×10^3 mm Hg?

5.9 The equilibrium constant for a certain reaction is 2.5×10^{-3} at $300°K$; ΔH for the reaction is +20.0 kcal. Using Equation 4, Table 5.3, calculate K at $400°K$.

5.10 Using Equation 5, Table 5.3, calculate the activation energy for a reaction where the rate constant doubles when the temperature increases from $27°C$ to $37°C$.

5.11 The potential, E, for the reduction of Zn^{2+} is given by the equation:

$$E = -0.76 - 0.030 \log [1/(\text{conc. } Zn^{2+})]$$

a. Calculate E when the concentration of Zn^{2+} is 1.0×10^{-8}.

b. Calculate the concentration of Zn^{2+} when $E = -1.52$.

5.12 The fraction, f, of molecules having an energy equal to or greater than the activation energy, E_a, is given by the equation:

$$f = e^{-E_a/RT}$$

What is f when $E_a = 1.00 \times 10^4$ cal., $T = 298°K$? ($R = 1.99$ cal/$°K$).

5.13 The Boltzmann equation for the distribution of molecules among two energy levels is:

$$\frac{n_2}{n_1} = e^{(E_1 - E_2)/RT}$$

where n_2 and n_1 are the numbers of molecules in the levels whose energies are E_2 and E_1, respectively. Calculate the ratio n_2/n_1 when $E_1 = 0$ and

a. $E_2 = 0$, T = $300°K$.

b. $E_2 = 1000$ cal., T = $300°K$.

c. $E_2 = 1000$ cal., T = $600°K$.

5.14 The world-wide consumption of energy is increasing at a rate of about 3 per cent per year.

a. If this increase continues, how long will it take to double our present rate of consumption?

b. About half of this 3 per cent increase can be attributed to an increase in population. If world population stabilized at the present level, how long would it take to double our present rate of energy consumption?

THE SLIDE RULE

We saw in Chapter 5 how a table of logarithms can be used to multiply, divide, raise to a power, or extract a root. These operations can be carried out more rapidly, with some loss of accuracy, by means of a slide rule. A simple 10-inch rule, carrying A, B, C, D, K, and L scales and selling for about $2, is adequate for most of the calculations involved in general chemistry. The time saved on home-work assignments, laboratory calculations, and examinations is well worth the cost of the rule.

To understand the principle upon which the slide rule is based, refer to Fig-ure 6.1, which shows the three scales C, D and L at the bottom of a simple rule. Notice that on the L scale, the numbers are spaced at equal intervals, much like those on an ordinary ruler. The numbers on the L scale represent the logarithms of those on the C and D scales (these latter two scales are identical except that the C scale is located on the movable slide while the D scale is on the stationary part of

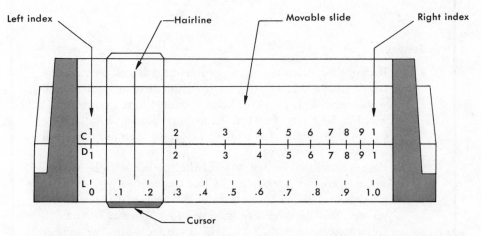

FIGURE 6.1 The positions of the numbers on the C and D scales are determined by their logarithms (L scale).

the rule). Notice, for example, that the number 2 (log 2 = 0.3010) on the D scale is located directly above 0.3 on the L scale; 4 on the D scale (log 4 = 0.6021) is directly above 0.6 on the L scale. In other words, the distances between numbers on the C and D scales are directly proportional to their logarithms. The number 2 is located about 3/10 of the way from the *left index* (1) to the *right index* (10); the number 4 about 6/10 of this distance, etc.

This design makes it possible to use the C and D scales to perform the operations of multiplication and division. The general principle is a very simple one: to multiply one number by another, we add distances which are proportional to the logarithms of these numbers. To divide one number by another, we subtract corresponding distances. The details of these operations are discussed in Section 6.2; the important point to note here is that a slide rule is simply a device which enables us to add and subtract logarithms mechanically. The incorporation of other scales (A, B, K · · ·) increases the number of arithmetical operations that can be performed (Sections 6.4, 6.5).

6.1 LOCATING NUMBERS

To illustrate how numbers can be located on the various scales of the slide rule, let us examine the D scale. The large digits (1, 2, 3, 4 · · ·) represent the integers between 1 and 10. Note that the spaces between these integers become smaller as we move from left to right. This gradual compression of the scale is a consequence of its logarithmic nature. The distance between the numbers 1 and 2 is about 3 inches (log 2 – log 1 = 0.3010), while that between 9 and 10 is less than half an inch (log 10 – log 9 = 0.0458).

We shall now consider how to find three-digit numbers in various positions of the D scale (Example 6.1). The positions of these numbers are indicated by dotted vertical lines in Figure 6.2.

Example 6.1 Locate the following numbers on the D scale:
- a. 1.34
- b. 3.42
- c. 6.53

Solution

- a. Focusing our attention on the portion of the D scale between the large digits 1 and 2, we recognize ten major divisions, each labeled with a small digit (1, 2, 3, 4) corresponding to the numbers 1.1, 1.2, 1.3, 1.4 · · · . Between the numbers 1.3 and 1.4, there are ten small divisions. The number 1.34 must then fall four small divisions to the right of 1.3.
- b. Here, we note that the ten major divisions between the large digits 3 and 4 are not numbered. To find the number 3.4, we move four major divisions to the right of the digit 3. To locate 3.42, we note that there are five small divisions between 3.4 and 3.5, corresponding to 3.42, 3.44, 3.46, 3.48, and 3.50. The number 3.42 must then be located one small division to the right of 3.4.

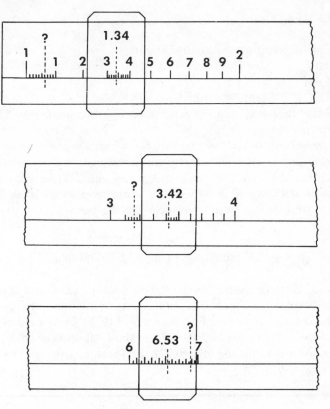

FIGURE 6.2 Location of the numbers 1.34, 3.42, and 6.53 on the D (or C) scale.

c. We first locate the number 6.5, five major divisions to the right of 6. Between 6.5 and 6.6, there is one small division, which must correspond to 6.55. The number 6.53 is located a little more than half way (viz., 3/5 of the way) between 6.50 and 6.55.

Example 6.1 and the accompanying discussion should enable you to locate any number between 1 and 10 on the C or D scale. But, you may well ask, where do I find a number such as 0.134, which is less than 1, or 13.4, which is greater than 10? You may be surprised to learn that you have already located these numbers! The slide rule setting corresponding to 1.34 also represents 0.134, 13.4, and indeed any number which can be expressed as 1.34×10^n where n is a whole number $(-1, 0, +1, \cdots)$. Similarly, the setting 3.42 locates numbers such as 0.00342 or 3420 (i.e., 3.42×10^{-3}, 3.42×10^3). In other words, *the location of a number on the slide rule is independent of the position of the decimal point or the power of 10 used to express the number in exponential notation.* This means, of course, that in carrying out any operation with the slide rule, the position of the decimal point must be determined independently. We will discuss a general method of doing this in Section 6.3, after we have shown how the slide rule is used to carry out some simple multiplications and divisions.

EXERCISES

1. Set the following numbers on the D scale.
   a. 2.02   d. 0.369
   b. 1.68   e. 51.5
   c. 3.54   f. 5.74 × 10^{-22}
2. Give the three-digit numbers corresponding to the unnumbered dotted lines in Figure 6.2.
3. It is reasonable to suppose that the error which a beginner makes in setting a number on the slide rule is equivalent to the smallest division on the D scale, which could be the distance between 1.99 and 2.00 or, alternatively, that between 9.95 and 10.00. (Note that these divisions are of about the same size.) What percentage of error does this represent?

6.2 MULTIPLICATION AND DIVISION

The basic principle involved in multiplication using the C and D scales of the slide rule is illustrated in Figure 6.3. Notice that to multiply 2 × 3, we add a distance on the D scale proportional to the log of 2 (0.3010) to a distance on the C scale proportional to the log of 3 (0.4771). The total distance covered on the D scale is proportional to the log of 6 (0.7781). The application of this principle is shown in Example 6.2.

Example 6.2 Multiply:
   a. 2.12 × 4.35
   b. 5.05 × 4.33

Solution

   a. We first set the 1 at the left of the C scale (the left index) directly above 2.12 on the D scale. Now, move the transparent plastic *cursor* so that its hairline lines up exactly with 4.35 on the C scale. Read the answer, approximately 9.22, at the hairline on the D scale.
   b. If we attempt to repeat the procedure of part (a), we find that the number 4.33 on the C scale falls beyond the end of the D

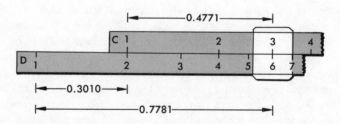

FIGURE 6.3 Multiplication on the slide rule is equivalent to adding logarithms: log 2 + log 3 = log 6.

(a)

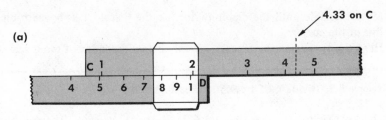

(b)

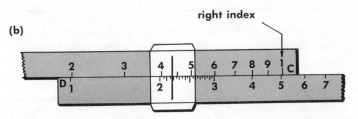

FIGURE 6.4 Multiplication (*b* diagram): 5.05 x 4.33 = 21.9. Note from diagram *a* that this multiplication cannot be achieved by setting the *left* index of the C scale above 5.05 on the D scale.

scale. (Figure 6.4) To get around this difficulty, we set the *right index* of the C scale above 5.05 on the D scale. Now, move the cursor so that its hairline coincides with 4.33 on the C scale. Directly beneath, on the D scale, we find that the hairline falls about half way across the space between 2.18 and 2.20. We might estimate its position to be 2.19. Knowing that the product must be approximately 20 (5 X 4 = 20), we take 21.9 to be our answer.

In general, to multiply one number by another, we:
1. **Set the appropriate index of the C scale (left or right) directly above the first number on the D scale.**
2. **Move the cursor so that its hairline falls on the second number, on the C scale.**
3. **Read the answer directly beneath the hairline on the D scale.**

Since the process of division is in the inverse of multiplication, it can be accomplished on the slide rule by reversing the operations involved in multiplication. Referring back to Figure 6.3, we note that by *subtracting* a distance proportional to the log of 3 (0.4771) on the C scale from a distance proportional to the log of 6 (0.7781) on the D scale, we obtain a distance proportional to the log of 2 (0.3010) on the D scale. In this way, we demonstrate that 6/2 = 3.

In general, to divide one number by another, we:
1. **Move the cursor so that its hairline falls directly over the numerator on the D scale.**

2. Move the slide until the denominator, on the C scale, falls beneath the hairline of the cursor.

3. Read the answer on the D scale, directly below the index of the C scale.

Example 6.3 Divide 6.05 by 9.50. (See Figure 6.5.)

Solution Using the cursor, line up 6.05 on the D scale with 9.50 on the C scale. Directly below the right index of the C scale, read the number "6.37" on the D scale. Remember that the slide rule does not locate the decimal point. Noting that 6.05/9.50 is somewhat less than 1, we deduce that the quotient must be 0.637 rather than 6.37.

Many times, in working problems, we are required to carry out a series of successive multiplications and divisions. In doing this, considerable time can be saved by alternating the processes, first dividing, then multiplying, then dividing, and so on.

Example 6.4 $\dfrac{(6.02)(1.88)(3.14)}{(1.24)(5.10)} = ?$

Solution Let us first divide 6.02 by 1.24, then multiply by 1.88, divide by 5.10, and, finally, multiply by 3.14.

    a. To divide 6.02 by 1.24, use the cursor to line up 6.02 on the D scale with 1.24 on the C scale. The quotient, which need not be recorded, can be found on the D scale below the left index of the C scale. Note that the rule is now in position to carry out a multiplication.

    b. To multiply by 1.88, move the cursor so that its hairline falls at 1.88 on the C scale. The product will appear on the D scale, beneath the hairline.

    c. To divide by 5.10, leave the cursor where it is and move the slide until 5.10 on the C scale lines up with the hairline.

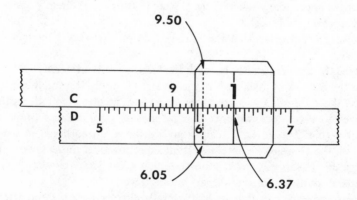

FIGURE 6.5 Division: 6.05/9.50 = 0.637.

d. To multiply by 3.14, move the cursor so that the hairline falls at 3.14 on the C scale. Read "5.62" on the D scale.

To locate the decimal point, we round off the numbers involved to one digit.

$$\frac{6 \times 2 \times 3}{1 \times 5} = \frac{36}{5} \approx 7$$

This tells us that the number must lie between 1 and 10 and hence must be 5.62 rather than 0.562 or 56.2.

Try carrying out all the multiplications first and then performing the divisions! You will find that several extra manipulations are involved. This should convince you that the recommended procedure of alternating multiplications with divisions has considerable merit.

EXERCISES

Perform the indicated operations on the slide rule.
1. 6.41×1.39
2. 8.54×2.90
3. $7.42/6.10$
4. $1.28/5.90$
5. $(6.19)(3.20)/(2.91)$
6. $(1.86)(3.95)/(2.87)(4.19)$
7. $(5.82)(4.19)(3.14)/(1.88)(6.12)$

6.3 LOCATING THE DECIMAL POINT (EXPONENT OF 10)

In Examples 6.2 through 6.4, we described what amounts to a "common sense" approach to locating the decimal point in a slide rule calculation. This involves rounding off all numbers to one digit so that a simple calculation will tell us whether the answer lies in the range 0.1–1, 1–10, 10–100, etc. In the cases cited, all of the numbers that entered into the calculations fell between 1 and 10, which simplifies matters considerably. However, we can readily extend this approach to any multiplication, division, or combined operation by adding a preliminary step. Before carrying out a slide rule calculation, **the numbers involved should first be expressed in exponential notation.** By applying the rules for the use of exponents (Chapter 4), we can then readily locate the decimal point or, if the answer is to be expressed in exponential notation, the appropriate power of 10. Suppose, for example, we wish to carry out the multiplication:

$$(0.0603)(1960) = ?$$

Rewriting the two numbers in exponential notation:

$$(6.03 \times 10^{-2})(1.96 \times 10^{3})$$

Following the rules for multiplication using exponents:

$$(6.03 \times 1.96) \times 10^1$$

On a slide rule, we find that the product is "118." Noting that

$$6.03 \times 1.96 \approx 6 \times 2 = 12$$

we see that the answer must be $11.8 \times 10^1 = 1.18 \times 10^2 = 118$

Example 6.5 applies the same approach to a somewhat more involved operation.

Example 6.5 With the aid of a slide rule, evaluate:

$$\frac{(0.0624)(192)}{(9.50 \times 10^{-4})(50.4)}$$

Solution The first step is to express each number in exponential notation:

$$\frac{(6.24 \times 10^{-2})(1.92 \times 10^2)}{(9.50 \times 10^{-4})(5.04 \times 10^1)}$$

Now, combining exponents, we obtain:

$$\frac{(6.24)(1.92) \times 10^0}{(9.50)(5.04) \times 10^{-3}} = \frac{6.24 \times 1.92}{9.50 \times 5.04} \times 10^3$$

The quotient $(6.24)(1.92)/(9.50)(5.04)$ is evaluated on the slide rule, as in Example 6.4, giving (in this case) "250." Noting that $(6 \times 2)/(10 \times 5) \approx 0.2$, we deduce that to 3 significant figures:

$$\frac{6.24 \times 1.92}{9.50 \times 5.04} = 0.250$$

The answer must therefore be 0.250×10^3 or 2.50×10^2.

EXERCISES

The following operations give the indicated slide rule readings. Express the answers in exponential notation.

Operation	Slide Rule Reading	Answer
1. 0.0249×0.352	876	8.76×10^{-3}
2. $(2.52 \times 10^3)/0.0630$	400	
3. $(3.49 \times 10^7)(0.519)(52.4)$	949	
4. $(9.60)(125)/0.0516$	233	
5. $\dfrac{(62.3)(0.817)}{(7.12)(1.57 \times 10^{-4})}$	455	

6.4 SQUARES AND SQUARE ROOTS

To square a number or extract its square root, we make use of the A and D scales. You will note that the A scale is made up of two ranges, each running from 1 to 1. It is convenient to think of the index at the center of the A scale as representing the number 10 and that at the far right as representing the number 100. From this point of view, the range at the left runs from 1 to 10, while that at the right runs from 10 to 100.

Squaring a number on the slide rule is a very simple procedure. All we do is to **set the hairline of the cursor above the number on the D scale and read its square on the A scale.** Thus, we find (Figure 6.6) that $2^2 = 4$, $3^2 = 9$, $(2.14)^2 = 4.58$. To find $(2.14 \times 10^{-6})^2$, we use the rules of exponents (Chapter 4).

$$(2.14 \times 10^{-6})^2 = (2.14)^2 \times (10^{-6})^2 = 4.58 \times 10^{-12}$$

Extracting the square root of a number is a slightly more complicated manipulation. In principle, all we have to do is to reverse the process involved in squaring the number. That is, we **set the hairline of the cursor above the number on the A scale and read the square root on the D scale.** In this way, we deduce that:

$$(4)^{1/2} = 2$$

$$(12.0)^{1/2} = 3.46$$

$$(50.0)^{1/2} = 7.07$$

A difficulty arises if the number whose square root we are asked to find does not fall between 1 and 100. In this case, **we rewrite the number in exponential form so that the coefficient is a number between 1 and 100 and the exponent is an even power** (e.g., $-2, 0, 2, 4 \cdots$).

> **Example 6.6** Find:
>   a. $(12,600)^{1/2}$
>   b. $(1.86 \times 10^5)^{1/2}$

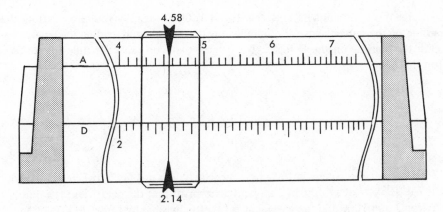

FIGURE 6.6 Squares of numbers: $(2.14)^2 = 4.58$.

Solution

a. $(12,600)^{1/2} = (1.26 \times 10^4)^{1/2} = (1.26)^{1/2} \times 10^2$

Setting the hairline of the cursor at 1.26 on the A scale, we read 1.12 on the D scale. Our answer is 1.12×10^2, or 112.

b. Here, we must first convert the exponent to an even number. To do this, we multiply 1.86 by 10 and divide 10^5 by 10:

$$(1.86 \times 10^5)^{1/2} = (18.6 \times 10^4)^{1/2} = (18.6)^{1/2} \times 10^2$$

To find the square root of 18.6, we move the hairline of the cursor to 18.6 on the A scale (note that 18.6 falls beyond 10, on the right half of the A scale). We read 4.31 on the D scale; our answer must then be 4.31×10^2.

EXERCISES

Find:

1. $(1.69)^2$
2. $(6.12 \times 10^{-3})^2$
3. $(1.60)^{1/2}$
4. $(87.0)^{1/2}$

5. $(2.02 \times 10^2)^{1/2}$
6. $(3.86 \times 10^7)^{1/2}$
7. $(2.98 \times 10^{-5})^{1/2}$

6.5 CUBES AND CUBE ROOTS

To cube a number or extract its cube root, we use the K scale in much the same way that the A scale is used to obtain squares or square roots. Notice that the K scale is divided into three segments, which may be thought of as running from 1 to 10 (left), 10 to 100 (center), and 100 to 1000 (right). To find the cube of a number, we **set the hairline of the cursor over that number on the D scale and read the answer under the hairline on the K scale.** Thus, we find that:

$$2^3 = 8; \quad (3.0)^3 = 27; \quad (8.00)^3 = 512$$
$$(3.0 \times 10^{-3})^3 = (3.0)^3 \times (10^{-3})^3 = 27 \times 10^{-9}$$

Cube roots of numbers between 1 and 1000 are readily found by **setting the hairline of the cursor above the number on the K scale and reading the answer beneath the hairline on the D scale.** We suggest that you use your slide rule to confirm that:

$$(3.00)^{1/3} = 1.44$$
$$(30.0)^{1/3} = 3.11$$
$$(300)^{1/3} = 6.69$$

To extract the cube root of a number which does not fall between 1 and 1000, it is necessary to rewrite it in exponential form so that the coefficient is a number between 1 and 1000 and the exponent is exactly divisible by three.

Example 6.7 Find:
  a. $(0.00190)^{1/3}$
  b. $(4.68 \times 10^{-5})^{1/3}$

Solution
  a. $(0.00190)^{1/3} = (1.90 \times 10^{-3})^{1/3} = (1.90)^{1/3} \times 10^{-1}$
  Below 1.90 on the K scale, we read 1.24 on the D scale. Our answer must be 1.24×10^{-1}, or 0.124.
  b. We must first rewrite the number so that the exponent, upon division by 3, will give a whole number. To obtain the desired form, we multiply the coefficient by 10 and divide the exponent by 10.

$$(4.68 \times 10^{-5})^{1/3} = (46.8 \times 10^{-6})^{1/3} = (46.8)^{1/3} \times 10^{-2}$$

Setting 46.8 on the K scale (between the second 4 and 5), we read its cube root on the D scale as 3.60. Hence, our answer must be 3.60×10^{-2}.

EXERCISES

1. $(6.08)^3$
2. $(1.40 \times 10^2)^3$
3. $(5.95 \times 10^{-4})^3$
4. $(12.9)^{1/3}$

5. $(268)^{1/3}$
6. $(1.96)^{1/3}$
7. $(1.37 \times 10^4)^{1/3}$
8. $(5.92 \times 10^{-4})^{1/3}$

6.6 LOGARITHMS

The only linear scale on the slide rule, where the numbers are spaced at equal intervals, is the L scale. This scale gives the logarithms of the numbers between 1 and 10. To find the logarithm of a number in this range, we set the hairline of the cursor above the number on the D scale and read its logarithm on the L scale. In this way, we find that the logarithm of 2 is, to three digits, 0.301, that of 3 is 0.477, and so on (Figure 6.7).

If the number whose logarithm we require is less than 1 or greater than 10, we proceed as described in Chapter 5. That is, we write the number in standard exponential notation, use the L scale to find the mantissa, and take the exponent to be the characteristic.

Example 6.8 Find the logarithm of 3.70×10^{-5}

Solution $\log(3.70 \times 10^{-5}) = \log 3.70 + \log 10^{-5} = \log 3.70 - 5$. Setting 3.70 on the D scale, we read its logarithm, on the L scale, to be 0.568. Hence, $\log(3.70 \times 10^{-5}) = 0.568 - 5 = -4.432$.

We see from this example that the L scale is equivalent to a three-place table of logarithms. All of the operations which were described in Chapter 5 in connection

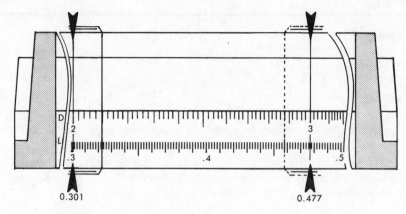

FIGURE 6.7   Logarithms of numbers: log 2 = 0.301; log 3 = 0.477.

with a table of logarithms can be carried out on a slide rule, using the L and D scales. We can, for example, use it to find antilogarithms.

Example 6.9   Find the numbers whose logarithms are:
   a.   0.380
   b.   3.100
   c. -2.900

Solution

   a.   Set the hairline of the cursor over 0.380 on the L scale; read the number, 2.40, on the D scale.
   b.   Set the hairline over the mantissa, 0.100, on the L scale. The number whose log is 0.100 is read from the D scale to be 1.26. The number whose log is 3 is 10^3. Therefore:

$$\text{antilog } 3.100 = 1.26 \times 10^3$$

   c.   Firsf, rewrite the number (recall the discussion in Chapter 5) as:

$$0.100 - 3$$

$$\text{antilog } (0.100 - 3) = 1.26 \times 10^{-3}$$

EXERCISES

Find, using the slide rule:
1. log 6.14
2. $\log (1.29 \times 10^5)$
3. $\log (5.84 \times 10^{-4})$
4. ln 2.82
5. antilog 0.462
6. antilog 1.250
7. antilog (-5.400)

6.7 OTHER USES OF THE SLIDE RULE

The slide rule operations performed most frequently in general chemistry have been described in Sections 6.2 and 6.4 through 6.6. Depending on the number of scales available (and the price!), it is possible to use the slide rule to carry out a variety of other operations. Even the simplest slide rule has scales (labelled S and T) which allow us to obtain trigonometric functions (sine, tangent).

From time to time, we may have occasion to find the reciprocal of a number (the reciprocal of a number, x, is defined as $1/x$). This can, of course, be done by treating it as an ordinary division and making use of the C and D scales. Where a series of reciprocals is required, it is somewhat more convenient to make use of the C and C1 scales. Notice that the C1 scale amounts to an inverse C scale; the numbers on the C1 scale start at 10 on the far left and decrease to 1 on the far right. To find the reciprocal of a number, we set the hairline of the cursor over that number on the C scale and read its reciprocal at the hairline on the C1 scale. In this way, you can verify that:

$$1/2 = 0.500; \quad 1/6 = 0.167; \quad 1/3.16 = 0.316$$

The C1 scale can also be used in conjunction with the D scale to carry out the processes of multiplication and division. In both cases, the manipulation is somewhat more rapid than is the case when the C and D scales are used (Section 6.2).

To multiply 9 by 12 using the C1 and D scales, use the cursor to line up 9 on the C1 scale with 12 on the D scale. Read the product, 108, on the D scale, directly below the index of the C1 (or C) scale. Note that when multiplications are carried out in this way, it is never necessary to shift indices to bring the answer onto the scale.

To divide 7.0 by any desired number (e.g., 5.0, 3.5, 2.0), set the right index of the C1 scale to 7 on the D scale. Move the cursor to the number by which you wish to divide, on the C1 scale. Read the answer at the hairline on the D scale. In this way, we find that:

$$7.0/5.0 = 1.4$$
$$7.0/3.5 = 2.0$$
$$7.0/2.0 = 3.5$$

Note the advantage of this method of performing divisions when you are asked to divide a given number, such as 7.0, by a series of numbers.

6.8 ELECTRONIC CALCULATORS

Within the past few years, several companies have put on the market simple electronic calculators designed to carry out the most common arithmetical operations of addition, subtraction, multiplication, and division. The most popular models, one of which is shown in Figure 6.8a, are pocket size ($\sim3'' \times 5''$) and sell in a price range of $60 to $130. If you are interested in buying a calculator of this type, you should check the review articles that have appeared in consumer maga-

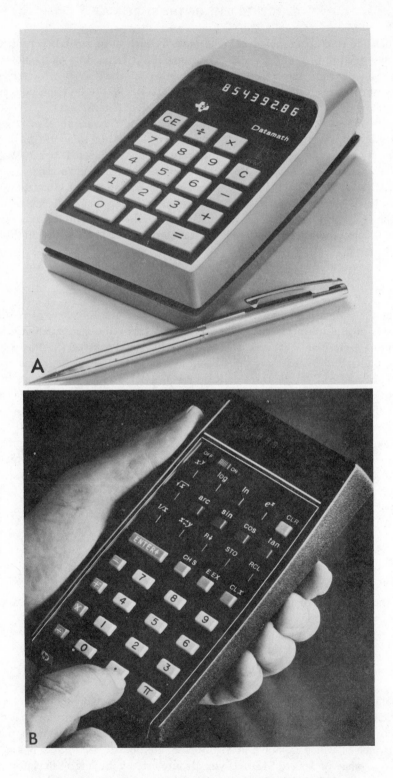

FIGURE 6.8   Two models of pocket-size electronic calculators.

zines (Consumer Reports, June 1973; Consumer Bulletin, May 1973) for a comparison of the merits of different brands.

The obvious advantage of a calculator over a slide rule is simplicity of operation. In a few minutes, you can learn to perform the basic operations on a calculator; considerably more practice is required to become proficient with a slide rule. Another advantage is increased accuracy; answers on a calculator ordinarily appear to 6 or 8 digits. In practice, this factor is of minor importance in general chemistry; very rarely will you need to carry out calculations beyond the 3 digits that can be obtained with a slide rule.

Before spending in the vicinity of $100 for an electronic calculator as opposed to as little as $2 for a slide rule, you should be aware of their limitations. Calculators in this price range are nowhere near as versatile as the simplest slide rule. They can be used for addition, subtraction, multiplication, and division, and essentially nothing else. You can not use them to take logarithms or square roots, which you will need from time to time in general chemistry. Combined operations of multiplication and division are awkward to carry out with many calculators. Indeed, a person skilled in the use of the slide rule can perform almost any operation (except addition and subtraction) as rapidly as a person operating a calculator.

It is possible to buy electronic calculators which will perform all the operations that can be carried out on a slide rule (plus addition and subtraction). One such instrument, which currently sells for about $300, is shown in Figure 6.8b. Like its less expensive and less versatile cousins, it is miniature in size, completely portable (rechargable batteries), and easy to understand and operate.

PROBLEMS

The following problems are designed to review material on exponents (Chapter 4) and logarithms (Chapter 5), in addition to improving your speed and accuracy in using the slide rule.

Make use of the slide rule to obtain numerical answers to each problem.

6.1 $(5.49 \times 10^2)(6.02 \times 10^{-8})$

6.2 $\dfrac{4.91 \times 10^{-2}}{6.84 \times 10^3}$

6.3 $(3.08 \times 10^7)^2$

6.4 $\log(2.41 \times 10^5)$

6.5 $(7.28 \times 10^{-6})^{1/3}$

6.6 antilog 0.649

6.7 $(5.00 \times 10^2)^{1/2}$

6.8 $(1.29 \times 10^{-22})^3$

6.9 $1/(1.98 \times 10^{-4})$

6.10 $\dfrac{(1.28 \times 10^7)(3.14)}{6.98 \times 10^3}$

6.11 $2.71(3.08 \times 10^{-3})^{1/2}$

6.12 $\dfrac{(6.47 \times 10^{-2})(8.19 \times 10^7)}{(3.25 \times 10^4)(4.21 \times 10^{-6})}$

6.13 antilog 2.490

6.14 3.23 log 6.00

6.15 $\dfrac{(6.19 \times 10^4)(2.00 \times 10^{-3})^2}{9.19 \times 10^4}$

6.16 $\dfrac{6.18(4.14 \times 10^{-5})^{1/3}}{2.00}$

6.17 $\dfrac{(9.62 \times 10^{-7})(4.08 \times 10^4)(1.72 \times 10^5)}{(4.43 \times 10^2)(5.09 \times 10^{-4})}$

6.18 $e^{1.80}$

6.19 1.26×10^3 (antilog 4.10)

6.20 $\dfrac{(2.19 \times 10^{-2})^{1/2}(4.80 \times 10^3)^2}{(6.18 \times 10^{-5})^{1/3}}$

CHAPTER 7

SIGNIFICANT FIGURES

The numbers that we deal with in general chemistry can be divided into two broad categories. Some numbers are, by their nature, **exact**. When we say that there are *two* crucibles or *five* beakers in a laboratory locker, we indicate the exact number of such items. Similarly, in the conversion factors:

$$1 \text{ ft} = 12 \text{ in}; \quad 1 \text{ liter} = 1000 \text{ ml}$$

the numbers 12 and 1000 are exact; by definition, there are precisely *twelve* inches in one foot and *one thousand* milliliters in one liter. Other numbers are **inexact**; when we refer to a "100 ml beaker," we do not imply that it has a volume of *exactly* 100 milliliters. If we fill the beaker with water, we may find that it holds as little as 90 ml or as much as 110 ml.

Numbers which arise from experimental measurements are always inexact. The uncertainty in a measurement depends upon the skill of the experimenter and the sensitivity of the instrument he uses. If we were asked to weigh out a sample of sodium chloride on a triple beam balance, we might be able to establish its weight to within 0.01 g. Perhaps we would report that it weighed:

$$2.65 \pm 0.01 \text{ g}$$

If we needed to know the mass more accurately, we could use a balance with a sensitivity of 0.001 g, in which case we might find that the sample weighed:

$$2.652 \pm 0.001 \text{ g}$$

Analytical balances capable of weighing to ±0.0001 g are readily available; using such a balance, we might report the mass to be:

$$2.6518 \pm 0.0001 \text{ g}$$

From the standpoint of a person who is trying to estimate the validity of an experiment or to repeat it in another laboratory, it is important that we specify the uncertainty associated with a measurement. One way to do this is to use the ± notation just shown. Frequently, we omit the ±0.01, ±0.001, and so on, and simply report:

2.65 g;   2.652 g;   2.6518 g

with the understanding that there is an *uncertainty of one unit in the last digit*. When we say that a sample of sodium chloride weighs "2.65 g," we imply that its mass is between 2.64 g and 2.66 g.

The precision of measurements such as this can also be described in terms of the number of **significant figures**. We say that in "2.65 g" there are three significant figures; each of the three digits is experimentally significant. The masses "2.652 g" and "2.6518 g" are quoted to 4 and 5 significant figures, respectively, implying successively higher degrees of precision.

We shall find the concept of significant figures a very useful one in expressing the reproducibility, not only of individual measurements, but also of calculated quantities based upon such measurements. In this chapter, we shall be concerned with the use of significant figures as a measure of experimental precision. Later, in Chapter 11, we will present a more sophisticated treatment of this topic based on statistical principles.

7.1 COUNTING SIGNIFICANT FIGURES

Frequently, we are faced with the problem of deciding how many significant figures there are in a number which arises from an experiment performed by another person. In many cases, there is no ambiguity. When we find in a table the atomic weight of calcium listed as 40.08, we trust that it is known to four significant figures.

When either the first or the last digit in an experimental quantity is zero, the number of significant figures may not be immediately obvious. Common sense is an invaluable guide here. When we find the atomic weight of krypton listed as 83.80, it should be clear that it is known to four significant figures. If the zero were not significant, there would be no reason for including it. Writing "83.80" implies that the true value of the atomic weight of krypton lies between 83.79 and 83.81.

In a slightly less obvious case, suppose we are told that the volume of a certain object is 0.02461 liters. Is the zero immediately to the right of the decimal point significant? A moment's reflection should convince you that it is not; in this case, *the zero is used simply to fix the position of the decimal point*. To make this conclusion more obvious, suppose we were to express the volume in milliliters rather than liters. Since 1 liter = 1000 ml, the volume would now be 24.61 ml, with four significant figures clearly indicated. Since we cannot change the precision of a measurement by changing the units in which it is reported, there must also be four significant figures in the quantity 0.02461 liters.

Suppose we are told to weigh out 500 g of sodium chloride for a certain experiment. To how many significant figures are we to make this measurement: one? two? three? Unfortunately, we cannot answer this question unless we are able to read the mind of the person who wrote the directions for the experiment. He might have meant to weigh out roughly a quantity between 400 and 600 g, i.e., 500 ± 100 g. If so, there is one significant figure. On the other hand, we may be expected to weigh to ±10 g (2 significant figures). Then again, perhaps we are supposed to weigh to the nearest gram, in which case the number "500" has three significant figures. About all we can do in cases like this is to wish that the person giving the directions had expressed the mass in standard exponential notation, such as:

$$5 \times 10^2 \text{ g} \quad \text{(1 significant figure)}$$

or: $\qquad\qquad\qquad\quad 5.0 \times 10^2 \text{ g} \quad \text{(2 significant figures)}$

or: $\qquad\qquad\qquad\quad 5.00 \times 10^2 \text{ g} \quad \text{(3 significant figures)}$

in which case there would have been no ambiguity.

If you think about it, the argument we have just presented leads us to a general rule for determining the number of significant figures in a cited quantity. We need merely to express that quantity properly in exponential notation (Chapter 4). **The number of digits in the coefficient of the power of 10 will equal the number of significant figures in the original quantity.** Thus, 0.0045630 kg, expressed in exponential notation, would be 4.5630×10^{-3} kg; the coefficient, 4.5630, has five digits, and hence, the original quantity contains five significant figures.

Example 7.1 How many significant figures are there in:

    a. 1.204×10^{-2} g $\qquad\qquad\qquad$ c. 0.00281 g

    b. 3.160×10^8 Å $\qquad\qquad\qquad$ d. 810 ml

Solution

    a. All digits are significant; 4.

    b. All digits are significant; 4.

    c. Three. The zeros serve only to fix the position of the decimal point (0.00281 g = 2.81×10^{-3} g).

    d. Perhaps 2 (8.1×10^2 ml) or 3 (8.10×10^2 ml).

Occasionally, you may be asked a question such as, "How many grams of carbon dioxide can be produced from one gram of carbon?" In a case such as this, where the number one is written out, you are to assume that it is exact. Essentially, you are being asked a rhetorical question, "How much carbon dioxide could be produced from one gram of carbon if the carbon could be weighed out exactly?" For calculation purposes, we assume an infinite number of significant figures in the number "one," just as we consider the numbers "1" and "12" in the defining equation:

$$1 \text{ ft} = 12 \text{ in}$$

to have an infinite number of significant figures.

EXERCISES

Give the number of significant figures in:

1.	12.82 liters	6.	0.0559 g
2.	3.19×10^{15} atoms	7.	2.92×10^2 g
3.	4.300×10^{-6} cm	8.	4.1 liters
4.	0.00641 g	9.	0.0002 cm
5.	8.2354×10^{-9} m	10.	450 g

7.2 MULTIPLICATION AND DIVISION

Most of the quantities that we measure in the laboratory are not end results in themselves; they are used to calculate other quantities. We may, for example, multiply the length of a piece of tin foil by its width to determine its area. In another case, we may divide the mass of a sample by its volume to determine its density.

When we multiply or divide two experimental quantities, both of which are inexact, the product or quotient is inexact. The question arises as to how great an error is introduced by these operations. To answer this question, it will be helpful to work through a specific example. Let us suppose that we have measured the mass and volume of a sample and found them to be 5.80 ±0.01 g and 2.6 ±0.1 ml, respectively. The density, found by dividing mass by volume, should be approximately (5.80/2.6) g/ml. However, taking into account the uncertainties in the experimental quantities, we realize that the density might be as large as (5.81/2.5) g/ml or as small as (5.79/2.7) g/ml. Carrying out these divisions, we obtain:

$$2.5 \overline{)5.81} \;= 2.32 \ldots \qquad 2.6 \overline{)5.80} \;= 2.23 \ldots \qquad 2.7 \overline{)5.79} \;= 2.14 \ldots$$

Comparing these three quotients, we conclude that the density is (2.2 ±0.1) g/ml. In other words, the density, like the volume, is known to 2 significant figures.

This example illustrates a general rule for the multiplication or division of inexact numbers. In general, we should **retain in the product or quotient the number of significant figures present in the least precise of the numbers.** Applying this rule, we deduce that:

$(6.10 \times 10^3)(2.08 \times 10^{-4})$	has	3 significant figures in answer
5.92×3.0	has	2 significant figures in answer
$8.2/3.194$	has	2 significant figures in answer

The preceding rule can save us considerable time in carrying out multiplications or divisions. Suppose we are asked to divide 8.2 by 3.194. Realizing that the answer can be given only to two significant figures, we might round off 3.194 to 3.2 before performing the division, thereby saving considerable time.*

---

*Many people prefer to retain one extra digit in carrying out a multiplication or division. In this case, they would round off 3.194 to 3.19 rather than 3.2, carry out the multiplication, and round off the answer to 2 significant figures. This procedure will change, at most, the last digit of the answer, perhaps by as much as 2 or 3 units.

Example 7.2 Carry out the following operations involving inexact numbers, retaining the correct number of significant figures:

    a. 6.19×2.8

    b. $3.18/1.702$

    c. $(4.10 \times 3.02 \times 10^9)/1.5$

Solution

    a. Since we are justified in retaining only two significant figures in our answer, we round off before multiplying and write:

$$6.2 \times 2.8 = 17$$

    b. $3.18/1.702 = 3.18/1.70 = 1.87$

    c. $(4.1 \times 3.0 \times 10^9)/1.5 = 8.2 \times 10^9$

Frequently, in carrying out operations of this sort, we use conversion factors which are either defined exactly (1 lb = 16 oz) or can be expressed to any desired number of digits (1 lb = 453.6 · · · g). The use of such conversion factors never changes the number of significant figures in our answer, the precision of which depends only upon those of the experimental quantities that enter into the calculation (Example 7.3).

Example 7.3 A student weighs a strip of magnesium on an analytical balance and finds its mass to be 0.106 g. Express this mass in pounds and ounces.

Solution To find the mass in pounds, we use the conversion factor 1 lb = 453.6 g:

$$0.106 \text{ g} \times \frac{1 \text{ lb}}{453.6 \text{ g}} = 2.34 \times 10^{-4} \text{ lb} \quad (3 \text{ significant figures})$$

Note that since the magnesium was weighed to only 3 significant figures, we could have rounded off the conversion factor to 3 figures, i.e., 1 lb = 454 g, before carrying out the calculation. To convert to ounces:

$$2.34 \times 10^{-4} \text{ lb} \times \frac{16 \text{ oz}}{1 \text{ lb}} = 3.74 \times 10^{-3} \text{ oz} \quad (3 \text{ significant figures})$$

We can use the rules governing significant figures to guide us in designing laboratory experiments. If we are to carry out two or more measurements and combine them by multiplying or dividing to obtain a final result, the precision of that result will be governed by that of the least precise measurement. Suppose we are asked to determine the density of a liquid, using a triple beam balance (±0.01 g) and a graduated cylinder (±0.1 ml). The precision of the density will be limited by the comparatively large uncertainty in the volume measurement. If we need to know the density more exactly, we must resort to a more precise device for measuring volumes, such as a pipet or volumetric flask (±0.01 ml). It would be fallacious to suppose that we could improve the precision by using an analytical balance capable of weighing to ±0.0001 g and then measuring volume to ±0.1 ml!

One final comment is in order concerning the rule that we have given for determining the number of significant figures in a product or a quotient. It is, at best, an approximation, albeit a convenient one. At times, it can lead to absurd situations. Let us suppose that two students, asked to determine the density of a liquid, obtain the following results:

	Mass	Volume	Density
Student 1	10.20 g	10.1 ml	1.01 g/ml
Student 2	10.10 g	9.9 ml	1.01 g/ml

Should the second student drop the last digit in his calculated density, reporting it as 1.0 g/ml, since there are only two significant figures in his volume? Common sense tells us that he should not. Like the other student, he has measured the volume to about ±1 per cent. It is quite reasonable for him to report his answer to ±1 per cent, i.e., as 1.01 g/ml rather than 1.0 g/ml, which would imply an uncertainty of ±10 percent.

Situations as extreme as this do not arise too frequently. They do, however, point out the limitations of using significant figures as a measure of precision and suggest that a more exact treatment of experimental errors would be in order (Chapter 11).

EXERCISES

Assuming that all the following numbers are inexact, carry out the indicated operations, giving answers to the correct number of significant figures.

1. $(2.49 \times 10^{-3})(3.81)$
2. 6.4023×19
3. 0.00481×212
4. $(3.18 \times 10^{-3})^2$

5. $7.17/6.2$
6. $8.73/5.198$
7. $\dfrac{6.48 \times 1.92}{5.2}$
8. $\dfrac{(8.10 \times 10^7)(4.43 \times 10^{-4})}{6.191 \times 10^2}$

7.3 ADDITION AND SUBTRACTION

When we add (or subtract) inexact numbers, we apply the general principle that the sum (or difference) **cannot have an absolute precision greater than that of the least precise number used in the computation.** To illustrate this point, suppose we add 1.32 g of sodium chloride and 0.006 g of potassium chloride to 28 g of water. How should we express the total mass of the resulting solution? The implied precisions of these masses are 0.01 g, 0.001 g, and 1 g, respectively.

sodium chloride:	1.32	± 0.01 g
potassium chloride:	0.006	± 0.001 g
water:	28	± 1 g
total mass:	29	± 1 g

The sum of the masses cannot be more precise than that of the water (± 1 g). We should write the total mass as 29 g, rather than 29.3 g, 29.326 g, or some other number.

In another case, suppose we weigh out a sample of sodium chloride by difference, starting with a mass of 32.241 g and ending with a mass of 32.13 g. The mass of the sample is:

$$
\begin{array}{r}
32.241 \text{ g} \\
- \ 32.13 \ \ \text{g} \\
\hline
0.11 \ \ \text{g}
\end{array}
$$

The mass is expressed only to the nearest hundredth of a gram, because that is the uncertainty in the final weighing.

From these illustrations, we see that in adding or subtracting experimental quantities, the principles are quite different from those governing multiplication and division. In particular, it is not true that the number of significant figures in a sum or difference is governed by the quantity having the fewest significant figures. In the first illustration, the mass of potassium chloride is known to only 1 significant figure, yet the sum can be expressed to 2. In the second case, the mass of sodium chloride found by difference is known to only 2 significant figures, even though the final and initial weighings carry 4 and 5, respectively. Cases such as these are by no means uncommon; we frequently "gain" significant figures in addition and "lose" them in subtraction. It is always the absolute precision (±1 g, ±0.01 g) which regulates the number of significant figures in a sum or difference.

Example 7.4 Carry out the following operations, being careful to give answers to the correct number of significant figures:
    a. 6.82 g + 2.111 g + 1268 g
    b. 213 g − 0.01 g
    c. 5.19×10^{-2} cc + 1.83 cc + 2.19×10^{2} cc

Solution

    a. Since "1268 g" has the lowest absolute precision (±1 g), the sum cannot have a greater precision. Therefore, we write:

$$7 \text{ g} + 2 \text{ g} + 1268 \text{ g} = 1277 \text{ g}$$

    b. The quantity 213 g is known only to ± 1 g, so subtracting 0.01 g from it has no effect. 213 g − 0 g = 213 g
    c. In order to obtain the sum, we first write out the numbers in decimal form:

$$5.19 \times 10^{-2} \text{ cc} = \quad 0.0519 \text{ cc} \pm 0.0001 \text{ cc}$$

$$1.83 \quad \text{cc} \pm 0.01 \text{ cc}$$

$$2.19 \times 10^{2} \quad \text{cc} = 219 \quad \text{cc} \pm 1 \text{ cc}$$

Since the sum cannot have a precision greater than 1 cc, we write:
$$0 + 2 \text{ cc} + 219 \text{ cc} = 221 \text{ cc}$$

EXERCISES

1. 9.10 g + 6.231 g = ?
2. 8.162 ml - 2.39 ml = ?
3. 4.30 cm + 29.1 cm + 0.345 cm = ?

4. 6.23 m + 915 cm - 12.7 cm = ?

5. 2.02×10^2 g - 9.6×10^1 g = ?
6. 3.18×10^{-1} ml + 1.6×10^{-2} ml = ?
7. (6.40 cm × 12.1 cm) - 2.19 cm² = ?
8. $\dfrac{3.18 \text{ lb}}{(1.92 \text{ in})(2.4 \text{ in})}$ - 0.17 lb/in² = ?

7.4 ROUNDING OFF NUMBERS

In calculations from experimental data, it is often necessary to drop one or more digits to obtain an answer with the appropriate number of significant figures. The rules which are followed in rounding off are:

1. If the first digit dropped is less than 5, leave the preceding digit unchanged (i.e., 3.123 → 3.12).
2. If the first digit dropped is greater than 5, increase the preceding digit by 1 (i.e., 3.127 → 3.13).
3. If the first digit dropped is 5, round off to make the preceding digit an even number* (i.e., 4.125 → 4.12; 4.135 → 4.14). The effect of this rule is that, on the average, the retained digit is increased half the time and left unchanged half the time.

Example 7.5 Round off each of the following to 3 significant figures:
   a. 6.167   c. 0.002245
   b. 2.132   d. 3135

Solution

   a. 6.17
   b. 2.13
   c. Since the digit to be retained, 4, is even, we leave it unchanged, giving 0.00224.
   d. Rounding off to an even number gives 3140.

EXERCISES

Round off 4.3154652 to the following numbers of significant figures:

1. 7
2. 6
3. 5
4. 4

5. 3
6. 2
7. 1

7.5 LOGARITHMS, ANTILOGARITHMS

Clearly, the precision of a number will govern the precision to be associated with its logarithm. This principle leads to the general rule that we should **retain in the**

---

*Left-handed people often prefer to round off to odd numbers. It really doesn't matter as long as you're consistent.

mantissa of the logarithm the same number of significant figures as there are in the number itself. Thus, we have:

$$\log 3.000 = 0.4771 \qquad \log 3.000 \times 10^5 = 5.4771$$

$$\log 3.00 = 0.477 \qquad \log 3.00 \times 10^3 = 3.477$$

$$\log 3.0 = 0.48 \qquad \log 3.0 \times 10^2 = 2.48$$

$$\log 3 = 0.5 \qquad \log 3 \times 10^{-4} = 0.5 - 4 = -3.5$$

Notice that, unless the characteristic is zero, there will be more digits in the logarithm than in the number itself. This is quite reasonable, since the characteristic serves only to fix the position of the decimal point or the power of 10.

When we are asked to find the number corresponding to a certain logarithm, we follow a rule entirely analogous to that stated above. **We retain in the antilogarithm the same number of significant figures that we have in the mantissa of the logarithm.**

$$\text{antilog } 0.3010 = 2.000 \qquad \text{antilog } 5.3010 = 2.000 \times 10^5$$

$$\text{antilog } 0.301 = 2.00 \qquad \text{antilog } 1.301 = 2.00 \times 10^1$$

$$\text{antilog } 0.30 = 2.0 \qquad \text{antilog } 2.30 = 2.0 \times 10^2$$

$$\text{antilog } 0.3 = 2 \qquad \text{antilog } (0.3 - 4) = 2 \times 10^{-4}$$

Example 7.6 Find, using a 4-place log table:
 a. log 6.19
 b. $\log (1.3 \times 10^{-4})$
 c. antilog 0.62

Solution

 a. Referring to the table of logarithms in Appendix 2, we find:

$$\log 6.190 = 0.7917$$

Hence, log 6.19 = 0.792 (3 significant figures)
 b. log 1.300 = 0.1139
  log 1.3 = 0.11
  $\log 1.3 \times 10^{-4} = 0.11 - 4 = -3.89$
 c. Scanning down the column labeled "0," we find that:

$$\log 4.100 = 0.6128$$

$$\log 4.200 = 0.6232$$

We deduce that the 2-digit number whose logarithm is closest to 0.62 is 4.2:

$$\log 4.2 = 0.62, \text{ or}$$

$$\text{antilog } 0.62 = 4.2$$

EXERCISES

Evaluate to the correct number of significant figures:

1. log 1.602
2. log 5.2
3. log 2.18×10^3
4. log 4.9×10^{-4}

5. antilog 2.0
6. antilog 0.185
7. antilog 3.20
8. antilog (-1.902)

7.6 THE SIGNIFICANCE OF SIGNIFICANT FIGURES

We have tried to point out throughout this chapter the importance of the proper use of significant figures in calculations based upon experimental data. A knowledge of the principles involved enables you to:

1. Communicate to another person the precision of your measurements (p. 91).
2. Design experiments in order to avoid unnecessary work (p. 93).
3. Save a considerable amount of time in calculations (p. 92).

On the other side of the coin, a student who habitually ignores the rules governing the use of significant figures creates the impression that he does not know what he is doing. Virtually every question that you are asked on homework assignments, quizzes, or examinations, describes the results of one or more experiments, usually of a quantitative nature. The nature of the question therefore requires that you express the answer to the correct number of significant figures. Most instructors will overlook an occasional lapse in this area. If, once in a while, you report one figure too many or one figure too few, you will probably get away with it. However, atrocities such as those committed by Students D and F in Figure 7.1 are unlikely to pass without comment. Any instructor reading these answers will conclude that the students haven't the foggiest notion of what experimental work is all about.

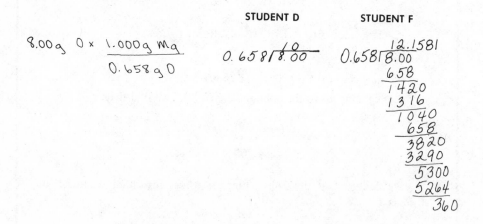

FIGURE 7.1   The use and misuse of significant figures.

PROBLEMS

Observe the rules of significant figures in expressing your answers.

7.1 A student finds that a sample of liquid weighing 12.562 g occupies a volume of 8.04 cc. Calculate the density of the liquid.

7.2 A block of metal with dimensions 5.2 cm X 2.1 cm X 4.6 cm weighs 109.82 g. Calculate the density of the metal.

7.3 A student weighs an empty flask, adds some potassium nitrate to it, and re-weighs. The two masses are 12.162 g and 12.498 g.
   a. What is the mass of potassium nitrate?
   b. What volume does the potassium nitrate occupy? (density = 2.107 g/ml)

7.4 Calculate the total mass of a solution prepared by adding 12.0 g of sodium chloride, 4.28 g of potassium nitrate, and 1.03 g of potassium chromate to 6.00×10^2 g of water.

7.5 In order to calibrate a pipet, a student weighs the water which drains from it after it has been filled to the mark. He obtains a mass of 9.9654 g. The density of water is 0.9970 g/ml. What is the volume of the pipet?

7.6 A sample of a certain compound weighing 2.040 g is found by analysis to contain 0.721 g of carbon and 0.050 g of hydrogen. It is known that the only other element present is iodine.
   a. What is the weight of iodine in the sample?
   b. What are the percentages by weight of the three elements in the compound?

7.7 When carbon burns in air, 3.667 g of carbon dioxide are formed for every gram of carbon. What mass of carbon dioxide is formed from samples of carbon weighing:
   a. 2.000 g
   b. 6.2 g

7.8 The molarity of a solution of sodium chloride can be expressed as:

$$M = \frac{\text{no. of grams NaCl}/58.44}{\text{no. of liters solution}}$$

A solution is prepared by dissolving 12.08 g of sodium chloride to give 9.00×10^2 ml. What is the molarity of this solution?

7.9 A sample of gas weighing 1.602 g occupies 224 ml at a pressure of 749 mm Hg and a temperature of 100.0°C. Calculate the molar weight (M) of the gas, using the equation:

$$M = \frac{gRT}{PV}$$

where g = mass in grams
T = temperature in °K = °C + 273.16
P = pressure in atm. = pressure in mm Hg/760.0
V = volume in ml
R = gas constant = 82.06 ml atm/mole °K

7.10 A student measures the pH of a certain solution, using instruments of successively greater precision. The values he obtains are as follows:
   a. 4
   b. 4.1
   c. 4.12
   d. 4.118
Using the definition: pH = -log(conc. H^+), calculate the concentration of H^+ corresponding to each measurement.

7.11 The equation relating the vapor pressure, P, of a liquid to the temperature can be written in the form:

$$\log P = \frac{-\Delta H}{2.303\ RT} + B$$

where R = 1.987 cal/°K, T = temperature in °K = °C + 273.16, ΔH = heat of vaporization in calories, and B is a constant known to be 5.169 for a particular liquid (when the vapor pressure is expressed in mm Hg).
   If the vapor pressure of the liquid is 12.8 mm Hg at 25.00°C, what is the numerical value of ΔH?

7.12 A certain element consists of two isotopes of masses 34.97 and 36.97. The atomic weight of the element can be calculated from the expression:

$$\text{A.W.} = x(36.97) + (1 - x)(34.97)$$

where x = mole fraction of the heavy isotope, which is known to be about 0.24. Express the atomic weight of the element to the correct number of significant figures.

ALGEBRAIC EQUATIONS

Many problems in chemistry are most readily solved by means of algebraic equations. Sometimes, all we have to do is to substitute directly into a fundamental equation such as the Ideal Gas Law, $PV = nRT$. More often, we find that such equations have to be manipulated to cast them in a form that we can use to calculate the desired quantity. Occasionally, it will be necessary to derive the equation we need from first principles.

In this chapter, we shall consider methods of setting up and solving the types of algebraic equations that are most frequently used in the first-year course in chemistry. These include equations in which the highest power to which an unknown is raised is 1 (first degree equation) or 2 (second degree equation).

8.1 FIRST DEGREE EQUATIONS IN ONE UNKNOWN

Equations of this type are readily solved by applying a fundamental principle of algebra which states that **an equation remains valid if the same operation is performed on both sides.** Specifically, we can:

1. *Add* the same quantity to both sides.
2. *Subtract* the same quantity from both sides.
3. *Multiply* both sides by the same quantity.
4. *Divide* both sides by the same quantity.

Several examples of first degree equations in one unknown, x, which can be solved by applying one or more of these rules are given in Table 8.1.

Notice that the general method of attack in all cases is to get all the terms involving the unknown, x, on one side of the equation, preferably in the numerator.

Many simple equations of fractional order $(1/2, 1/3, \cdots)$ or an order greater than 1 $(2, 3, \cdots)$ can be converted to first order equations by applying the principle that an equation remains valid if:

1. Both sides are raised to the same power.
2. The same root of each side is taken.

Examples of these two operations are given in Table 8.2.

Table 8.1   Solution of First Degree Equations in One Unknown

Equation	Operation
1. $5x = 4$ $x = 4/5 = 0.8$	Divide both sides by 5
2. $3.0 = \dfrac{0.063}{x}$ $3.0x = 0.063$ $x = 0.021$	Multiply both sides by x Divide both sides by 3.0
3. $3x = 4 - 5x$ $8x = 4$ $x = 4/8 = 0.5$	Add $5x$ to both sides Divide both sides by 8
4. $\dfrac{5}{2x - 1} = \dfrac{2}{x + 4}$ $5(x + 4) = 2(2x - 1)$ $5x + 20 = 4x - 2$ $x + 20 = -2$ $x = -22$	Multiply both sides by $(2x - 1)(x + 4)$ Subtract $4x$ from both sides Subtract 20 from both sides

Table 8.2   Conversion to First Degree Equations

Equation	Operation
1. $x^{1/2} = 6$ $x = 36$	Square both sides
2. $x^2 = 1.0 \times 10^{-12}$ $x = 1.0 \times 10^{-6}$	Extract square root of both sides
3. $4x^3 = 3.2 \times 10^{-11}$ $x^3 = 0.8 \times 10^{-11} = 8.0 \times 10^{-12}$ $x = 2.0 \times 10^{-4}$	Divide by 4 Extract cube root of both sides.

EXERCISES

1. Solve for x:
   a. $4x = 6.0 \times 10^{-4}$
   b. $(2x - 4)/6 = 3 - 4x$
   c. $1.98 = 2.64/x$
   d. $2/x - 1/6 = 1.0$

   e. $x^{1/3} = 2.2$
   f. $3x^2 = 13$
   g. $27x^4 = 1.0 \times 10^{-12}$
   h. $\dfrac{x^2}{(1 - x)^2} = 6.0$

2. Solve each equation for the quantity in bold type.
   a. $\dfrac{P_2}{P_1} = \dfrac{V_1}{V_2}$
   b. $PV = nRT$

   c. $E = mu^2/2$
   d. $X = \dfrac{m}{m + 1000/18.0}$

8.2   FIRST ORDER EQUATIONS IN CHEMISTRY

Many problems in general chemistry can be solved by substituting numbers directly into a first order algebraic equation and carrying out one or more of the simple operations described in Section 8.1. Examples 8.1 through 8.3 are typical of this type of problem.

Example 8.1 The relationship between temperature expressed in degrees Fahrenheit (°F) and degrees Celsius (°C) is:

$$°F = 1.8°C + 32°$$

    a. Convert 25°C to °F
    b. Convert 98.6°F (normal body temperature) to °C

Solution

    a. Here we can solve for °F by direct substitution.

$$°F = 1.8(25°) + 32° = 77°$$

    b. In this case, somewhat more algebra is involved. We might first solve the basic equation for the quantity we want, °C:

$$°F - 32° = 1.8°C \quad \text{(subtract 32° from each side)}$$

$$°C = \frac{°F - 32°}{1.8} \quad \text{(divide each side by 1.8)}$$

and then substitute numbers:

$$°C = \frac{98.6° - 32°}{1.8} = \frac{66.6°}{1.8} = 37.0°$$

    Alternatively, we could have started by substituting numbers into the basic equation:

$$98.6° = 1.8°C + 32°$$

and then solved for °C:

$$1.8°C = 98.6° - 32° = 66.6° \quad \text{(subtract 32° from both sides)}$$

$$°C = \frac{66.6°}{1.8} = 37.0° \quad \text{(divide both sides by 1.8)}$$

Example 8.2 The relationship between the pressure (P), volume (V), number of moles (n), and absolute temperature (T) of an ideal gas is expressed by the equation:

$$PV = nRT$$

where R, the gas constant, is 0.0821 lit atm/mole °K. What volume is occupied by 1.60 moles of an ideal gas at 1.20 atm and 300°K?

Solution We first solve the equation for the quantity we want, V. To do this, we divide both sides by P:

$$V = \frac{nRT}{P}$$

All the quantities on the right side of this equation are given in the statement of the problem; we can obtain V by direct substitution.

$$V = \frac{(1.60 \text{ moles}) \left(0.0821 \dfrac{\text{liter atm}}{\text{mole °K}}\right)(300°K)}{1.20 \text{ atm}} = 32.8 \text{ liters}$$

Example 8.3 The average velocity in cm/sec, u, of an oxygen molecule is given by the expression:

$$u = 2.79 \times 10^3 T^{1/2}$$

where T is the temperature in °K. At what temperature will an oxygen molecule have an average velocity of 5.00×10^4 cm/sec?

Solution To convert to a first order equation in T, we square both sides:

$$u^2 = (2.79 \times 10^3)^2 T = 7.79 \times 10^6 T$$

Solving for T and substituting numbers:

$$T = \frac{u^2}{7.79 \times 10^6} = \frac{(5.00 \times 10^4)^2}{7.79 \times 10^6} = \frac{25.0 \times 10^8}{7.79 \times 10^6} = 321°K$$

Most students have relatively little difficulty in setting up and solving problems such as those in Examples 8.1 to 8.3, where the appropriate algebraic equation is available and the substitution is more or less obvious. Occasionally, they may get bogged down in the algebra, but the analysis of the problem is straightforward.

Unfortunately, many problems in general chemistry which can, in principle, be solved by using a first order equation, do not unravel as simply as these examples. Sometimes, as in Example 8.4, you can readily write down the equation that applies (provided you know what molarity means!), but it may not be obvious where to go from there.

Example 8.4 What is the molarity of a solution prepared by dissolving 50.0 g of NaCl in 450 g of water? The density of the solution is 1.071 g/ml.

Solution The appropriate equation is clearly that which defines molarity:

$$\text{molarity} = \frac{\text{number of moles of solute (NaCl)}}{\text{number of liters of solution}} = \frac{n}{V}$$

Neither of the quantities we need, n or V, is directly available from the statement of the problem. However, the number of moles of NaCl is readily calculated from the number of grams, since we know that the molar weight of NaCl is 58.5 g.

$$n = \frac{50.0 \text{ g}}{58.5 \text{ g/mole}} = 0.855 \text{ mole}$$

At first glance, it is by no means obvious how we are to obtain V, the number of liters of solution. Notice, however, that we are given the density of the solution; the chances are we are supposed to use it! Realizing that density is defined as:

$$\text{density} = \frac{\text{mass}}{\text{volume}}$$

it should be clear that we could use the density given us to calculate the volume of the solution if we knew the total mass. Careful reading of the statement of the problem indicates that the total mass can be calculated by simply adding the weights of sodium chloride (50.0 g) and water (450 g).

If you can get this far in the analysis of the problem, the rest of the way is all downhill. Solving the density equation for volume and substituting numbers:

$$\text{volume} = \frac{\text{mass}}{\text{density}} = \frac{50 \text{ g} + 450 \text{ g}}{1.071 \text{ g/ml}} = 467 \text{ ml}$$

The quantity we need is the volume in liters:

$$V = \frac{467}{1000} \text{ liter} = 0.467 \text{ liter}$$

Now we can make the final substitution in the defining equation for molarity:

$$\text{molarity} = \frac{n}{V} = \frac{0.855 \text{ mole}}{0.467 \text{ liter}} = 1.83 \text{ mole/liter}$$

Occasionally, it is necessary to modify a basic equation to apply it to solve the problem at hand (Example 8.5).

Example 8.5 Using the ideal gas law, $PV = nRT$, calculate the molar weight of a gas if 1.10 g occupies a volume of 225 ml at 1.00 atm pressure and 373°K ($R = 0.0821$ lit atm/mole °K).

Solution We should be grateful to whoever phrased this problem for telling us what equation to use to solve it. However, the equation does not seem to be in the form we need to calculate the molar weight, M. Apparently, we are expected to manipulate the ideal gas law to obtain an expression for M. This procedure is readily accomplished provided we realize that the number of moles, n, can be expressed as the number of grams, g, divided by the molar weight, M (recall Example 8.4). That is:

$$n = \frac{g}{M}$$

Making this substitution in the ideal gas law, we obtain:

$$PV = \frac{gRT}{M}$$

Simple algebraic manipulation gives us the equation we need:

$$M = \frac{gRT}{PV}$$

All the required quantities are available; substitution gives:

$$M = \frac{(1.10 \text{ g}) \left(0.0821 \dfrac{\text{liter atm}}{\text{mole °K}}\right) (373°\text{K})}{(1.00 \text{ atm})(0.225 \text{ liter})} = 150 \text{ g/mole}$$

In certain general chemistry problems, you will be required to derive your own algebraic equation, drawing upon your knowledge of chemical principles (Example 8.6).

Example 8.6 Analysis of a mixture of SnO and SnO_2 weighing 1.000 g shows that it contains 0.850 g of tin. Calculate the percentage of SnO in the mixture.

Solution This problem differs from any considered previously in that it cannot be solved by substitution into a fundamental equation. You are "on your own" in the sense that you must derive the appropriate algebraic equation.

The key to unraveling this problem is to recognize that, since SnO and SnO_2 contain different percentages by weight of tin, the amount of tin in the sample must be related in a relatively simple way to its composition. To find out what this relationship is, we might start by finding out how much tin would be obtained (1) if the mixture were pure SnO (2) if the mixture were pure SnO_2.

1. If the mixture were pure SnO, there would be one mole of Sn (119 g) for every mole of SnO (135 g). Hence, a one gram sample would yield:

$$1.000 \text{ g} \times \frac{119 \text{ g}}{135 \text{ g}} = 0.881 \text{ g Sn}$$

2. If the mixture were pure SnO_2, there would be one mole of Sn (119 g) for every mole of SnO_2 (151 g). Hence, a one gram sample would yield:

$$1.000 \text{ g} \times \frac{119 \text{ g}}{151 \text{ g}} = 0.788 \text{ g Sn}$$

It is reassuring to find that the amount of tin actually obtained, 0.850 g, falls between these two extremes. This result suggests that we are on the right track. Indeed, since 0.850 is closer to 0.881 than it is to 0.788, we deduce that the mixture is mostly SnO. However "mostly SnO" would probably not be an acceptable answer.

To complete this problem, we need an algebraic equation relating the number of grams of tin obtained to the number of grams of SnO and SnO_2 in the sample. If we let the number of grams of SnO be x, the number of grams of SnO_2 must be $(1.000 - x)$, since the total weight of the sample is one gram. Now, for every gram of SnO in the sample, 0.881 g of Sn is produced. Hence, for x grams of SnO:

no. of grams Sn from SnO = 0.881 x

Similarly, since 0.788 g of Sn is formed for every gram of SnO_2, the amount of Sn formed from $(1.000 - x)$ g of SnO_2 must be:

no. of grams Sn from SnO_2 = 0.778(1.000 - x)

Knowing that the total weight of tin produced is 0.850 g, our final algebraic equation becomes:

$$0.850 = 0.881x + 0.778(1.000 - x)$$

which is readily solved to give:

$$x = \frac{0.850 - 0.778}{0.881 - 0.778} = \frac{0.072}{0.103} = 0.70$$

The percentage of SnO is:

$$\frac{\text{weight SnO}}{\text{total weight sample}} \times 100 = \frac{0.70 \text{ g}}{1.00 \text{ g}} \times 100 = 70\%$$

while that of SnO_2 is $(100 - 70) = 30\%$.

Perhaps the most discouraging feature of this problem is that the equation derived with so much effort will probably never be of use to us again. It is unlikely that anyone will be required to calculate the percentages of SnO and SnO_2 in a sample more than once in his life. However, the general line of analysis that we went through can be applied to a variety of problems (see, for example, Problem 8.8 at the end of this chapter).

You will probably agree that the examples just worked (especially Example 8.6) are not simple. The difficulty lies not in the algebra involved but in translating the word statement of the problem into an algebraic equation that can be solved for a numerical answer. If you could see how to do this immediately, without looking at the solutions, you are a big jump ahead of most beginning students. In each case, it was necessary to recognize a fact or principle that was by no means obvious. We have no "magic formula" that will enable you to come up with the "inspiration" that is the key to the solution of a problem of this type. It may be helpful, however, to reemphasize certain of the general rules of problem analysis discussed in Section 1.1 of Chapter 1. In order to solve any problem that is at all complex, you must:

1. Understand thoroughly the meaning of each term used in the statement of the problem. (The student who doesn't know what molarity and density mean is unlikely to get very far with Example 8.4.)

2. Know what the symbols used in an equation mean and how they may be expressed in terms of other quantities. (To work Example 8.5, you must realize that $n = g/M$.)

3. Have a broad enough knowledge of the principles of chemistry to choose the one that applies to the problem at hand. (In Example 8.6, it is essential that you recognize the relationship between the formula of a compound and its composition by weight.)

EXERCISE

1. When one gram of zinc reacts with HCl, 0.0153 mole of H_2 is produced. The reaction of one gram of aluminum with HCl gives 0.0556 mole of H_2. Analysis of an Al-Zn alloy shows that a sample weighing 0.400 grams gives 0.0170 mole of H_2. What are the percentages of Al and Zn in the alloy?

8.3 ALGEBRAIC EQUATIONS IN CHEMICAL EQUILIBRIA

In general chemistry, algebraic equations arise most frequently in problems dealing with chemical equilibria. It would not be appropriate here to discuss all of the principles that apply in such problems, but it will be necessary to discuss briefly the concept of the **equilibrium constant**. You should refer to your text for a more complete discussion of the significance of this quantity and its applications in chemistry.

The Law of Mass Action tells us that for a chemical system at equilibrium, there is an algebraic equation that relates the concentrations of the various species. For the general system:

$$aA + bB \rightleftharpoons cC + dD$$

where A, B, C, and D are species in solution (usually a gaseous or aqueous solution) and a, b, c, and d are the coefficients that appear in the balanced equations, the following equation applies:

$$\frac{[C]^c \times [D]^d}{[A]^a \times [B]^b} = K_c$$

where the square brackets indicate equilibrium concentrations and K_c, the equilibrium constant, is a number characteristic of a particular reaction at a given temperature.

To illustrate what this statement means, let us consider a system at equilibrium which contains the three species HI, H_2, and I_2. The balanced equation used to describe this system can be written:

$$2HI(g) \rightleftharpoons H_2(g) + I_2(g)$$

in which case, the expression for the equilibrium constant is:

$$\frac{[H_2] \times [I_2]}{[HI]^2} = K_c = 0.016 \text{ at } 520°C$$

Other examples are given in Table 8.3. Note that in each case, the concentrations of products (right side of equation) appear in the numerator; those of reactants (left side of equation) appear in the denominator. The power to which each concentration is raised is given by the coefficient of the species in the balanced equation.

Students commonly find problems involving equilibrium constants among the most challenging of those in the beginning course. In part, this is because such

Table 8.3 Examples of Equilibrium Constant Expressions

1. $2HI(g) \rightleftharpoons H_2(g) + I_2(g)$	$K_c = \dfrac{[H_2] \times [I_2]}{[HI]^2}$
2. $N_2(g) + O_2(g) \rightleftharpoons 2NO(g)$	$K_c = \dfrac{[NO]^2}{[N_2] \times [O_2]}$
3. $N_2(g) + 3H_2(g) \rightleftharpoons 2NH_3(g)$	$K_c = \dfrac{[NH_3]^2}{[N_2] \times [H_2]^3}$
4. $HC_2H_3O_2(aq) \rightleftharpoons H^+(aq) + C_2H_3O_2^-(aq)$	$K_a^* = \dfrac{[H^+] \times [C_2H_3O_2^-]}{[HC_2H_3O_2]}$

*The equilibrium constant for this type of reaction, the dissociation of a weak acid into ions, is referred to as the "dissociation constant" or "ionization constant" of the acid and is given the special symbol K_a.

problems come in what often seems to be a bewildering variety of different forms. Here, we will concentrate upon one aspect that is common to virtually all equilibrium problems. The operation that we will examine can be stated as follows:

Given the numerical value of K_c for a particular equilibrium system and the initial concentrations of all species, set up an algebraic equation in one unknown that can be solved to yield the equilibrium concentrations of all species.

In some cases, the algebraic equation obtained can be solved by reducing to a first order equation and applying the rules described in Section 8.1. More frequently, we obtain second or higher order equations which can be solved by methods to be discussed in Section 8.4.

Example 8.7 For the reaction that occurs when acetic acid dissociates in water:

$$HC_2H_3O_2(aq) \rightleftharpoons H^+(aq) + C_2H_3O_2^-(aq)$$

K_a is 1.8×10^{-5}. Set up an algebraic equation in one unknown which could be solved to give the concentrations of all species at equilibrium, starting with a concentration of acetic acid of 0.100 mole/liter.

Solution Let us choose our unknown, x, to be the equilibrium concentration of H^+ ion. Looking at the balanced equation, we see that for every mole of H^+ that is formed, one mole of $C_2H_3O_2^-$ is formed and one mole of $HC_2H_3O_2$ is consumed. Hence, if we form x moles/liter of H^+, we must form x moles/liter of $C_2H_3O_2^-$ and use up x moles/liter of $HC_2H_3O_2$. Since we started with no $C_2H_3O_2^-$, its concentration at equilibrium, like that of H^+, must be x. Starting with 0.100 mole/liter of $HC_2H_3O_2$ and consuming x moles/liter leaves us with $(0.100 - x)$ mole/liter at equilibrium. It is convenient to summarize our reasoning in the form of a table.

	Orig. Conc. (Mole/Liter)	Change	Equil. Conc. (Mole/Liter)
$HC_2H_3O_2$	0.100	$-x$	$0.100 - x$
H^+	0.000	$+x$	x
$C_2H_3O_2^-$	0.000	$+x$	x

Noting that the equilibrium constant expression has the form:

$$K_a = 1.8 \times 10^{-5} = \frac{[H^+] \times [C_2H_3O_2^-]}{[HC_2H_3O_2]}$$

we see that our algebraic equation becomes:

$$1.8 \times 10^{-5} = \frac{(x)(x)}{0.100 - x} = \frac{x^2}{0.100 - x}$$

This equation cannot be reduced to a first order equation by the methods described in Section 8.1. Methods of solving it will be discussed in Sections 8.4 and 8.5.

Example 8.8 The equilibrium constant for the reaction:

$$2HI(g) \rightleftharpoons H_2(g) + I_2(g)$$

is 0.010 at a particular temperature. Set up and, if possible, solve an algebraic equation which relates the equilibrium concentrations of all species, given
    a. orig. conc.: HI = 1.20 mole/liter, $H_2 = 0$, $I_2 = 0$
    b. orig. conc.: HI = 0, $H_2 = I_2 = 0.50$ mole/liter

Solution The equilibrium constant expression is, in both cases,

$$K_c = 0.010 = \frac{[H_2] \times [I_2]}{[HI]^2}$$

    a. Here, it will be convenient to let $x = [H_2]$. With this choice of variable, our table becomes:

	Orig. Conc. (Mole/Liter)	Change	Equil. Conc. (Mole/Liter)
H_2	0.00	$+x$	x
I_2	0.00	$+x$	x
HI	1.20	$-2x$	$1.20 - 2x$

Perhaps the most interesting feature of this table is the change in concentration of HI, which we have listed as $-2x$. The reasoning here is straightforward. The balanced equation tells us that *2* moles of HI are consumed for every mole of H_2 formed. Consequently, if the concentration of H_2 increases by x, that of HI must decrease by $2x$.
    The algebraic equation is:

$$\frac{x^2}{(1.20 - 2x)^2} = 0.010$$

As it happens, this equation can be solved by extracting the square root of both sides.

$$\frac{x}{1.20 - x} = (0.010)^{1/2} = 0.10$$

Simple algebra then allows us to find x.

$$x = 0.12 - 0.10x; \quad 1.10x = 0.12; \quad x = 0.11 \text{ mole/liter}$$

We deduce that the equilibrium concentrations of H_2, I_2, and HI must be 0.11, 0.11, and (1.20 - 0.22) = 0.98 mole/liter, respectively.
    b. Here, since there is no HI to start with, some of the H_2 and I_2 must react to bring the system to equilibrium. If we let x represent the

number of moles/liter of H_2 that is used up in this process, the table becomes:

	Orig. Conc. (Mole/Liter)	Change	Equil. Conc. (Mole/Liter)
H_2	0.50	$-x$	$0.50 - x$
I_2	0.50	$-x$	$0.50 - x$
HI	0.00	$+2x$	$2x$

Note that the changes in concentration of H_2 and I_2 are negative; these species are being consumed. The change in concentration of HI is positive, since it is formed when the reaction goes to equilibrium.

The algebraic equation is:

$$\frac{(0.50 - x)^2}{(2x)^2} = 0.010$$

which can be solved, as in part (a), by extracting the square root of both sides.

$$\frac{0.50 - x}{2x} = 0.10; \quad 0.50 - x = 0.20x; \quad 1.20x = 0.50; \quad x = 0.42$$

Referring back to the table, we see that $[H_2] = [I_2] = 0.50 - 0.42 = 0.08$ mole/liter, while $[HI] = 2(0.42) = 0.84$ mole/liter.

You may be curious about the way we chose the unknown in this part of the problem. Perhaps you would have preferred to work with the concentration of HI formed, letting that be y. If you did this, the table would be:

	Orig. Conc. (Mole/Liter)	Change	Equil. Conc. (Mole/Liter)
H_2	0.50	$-\frac{1}{2}y$	$0.50 - \frac{1}{2}y$
I_2	0.50	$-\frac{1}{2}y$	$0.50 - \frac{1}{2}y$
HI	0.00	$+y$	y

(Since 1 mole of H_2 yields 2 moles of HI, only $\frac{1}{2}y$ moles of H_2 are needed to form y moles of HI.) Setting up the equation and carrying out the algebra, we have:

$$\frac{(0.50 - \frac{1}{2}y)^2}{y^2} = 0.010; \quad \frac{0.50 - \frac{1}{2}y}{y} = 0.10$$

$$0.50 - \frac{1}{2}y = 0.10y; \quad 0.60y = 0.50; \quad y = 0.83; \quad \frac{1}{2}y = 0.42$$

$$[H_2] = 0.50 - 0.42 = 0.08; \quad [I_2] = 0.50 - 0.42 = 0.08; \quad [HI] = 0.83$$

The calculated concentrations are essentially identical with those obtained before. In general, in any equilibrium problem of this type, *it doesn't matter what you take to be the unknown, provided you are consistent in relating all the other concentrations to it.*

We have gone through these examples in considerable detail to illustrate the logic behind the derivation of algebraic equations to correspond to equilibrium constant expressions. You can practice this technique by working the following exercises, which illustrate the same principle with different systems.

EXERCISES

Given that:

$$HF(aq) \rightleftharpoons H^+(aq) + F^-(aq); \quad K_a = 7.0 \times 10^{-4}$$

$$2CO_2(g) \rightleftharpoons 2CO(g) + O_2(g); \quad K_c = 5.0 \times 10^{-4}$$

$$N_2(g) + 3H_2(g) \rightleftharpoons 2NH_3(g); \quad K_c = 5.0 \times 10^2$$

Complete the following "equilibrium tables," and write an algebraic equation which could be solved to give the equilibrium concentrations of all species.

	Orig. Conc. (Mole/Liter)	Change	Equil. Conc. (Mole/Liter)
1. HF	1.00	_____	_____
H$^+$	0.00	$+x$	_____
F$^-$	0.00	_____	_____
2. HF	0.50	$-x$	_____
H$^+$	0.10	_____	_____
F$^-$	0.00	_____	_____
3. CO$_2$	1.00	_____	_____
CO	0.00	_____	_____
O$_2$	0.00	$+x$	_____
4. CO$_2$	1.00	_____	_____
CO	0.00	_____	_____
O$_2$	1.00	$+x$	_____
5. N$_2$	1.00	$-x$	_____
H$_2$	3.00	_____	_____
NH$_3$	0.00	_____	_____
6. N$_2$	1.00	_____	_____
H$_2$	3.00	_____	_____
NH$_3$	0.00	$+x$	_____

8.4 SECOND DEGREE EQUATIONS IN ONE UNKNOWN

Equations of this type, sometimes referred to as *quadratic equations*, arise frequently in problems dealing with chemical systems at equilibrium (Examples 8.7, 8.8). Sometimes they can be solved by extracting the square root of both sides of the equation (Example 8.8). In many cases, however, this type of solution is not possible. Consider, for example, the equation:

$$\frac{x^2}{1 - x} = 4$$

Clearly, we cannot solve for x by extracting the square root of both sides of the equation.

Any second degree equation in one unknown can be solved by applying the so-called "quadratic formula." To use this approach, we rewrite the equation, if necessary, to get it in the form:

$$ax^2 + bx + c = 0 \qquad (8.1)$$

where a, b, and c are numbers. The two roots of this equation are:

$$x = \frac{-b \pm \sqrt[2]{b^2 - 4ac}}{2a} \qquad (8.2)$$

To illustrate how the quadratic formula is applied, let us use it to find the two values of x that satisfy the equation:

$$\frac{x^2}{1 - x} = 4$$

We first rewrite this equation to get it in the proper form:

$$x^2 = 4 - 4x$$

$$x^2 + 4x - 4 = 0$$

Comparing the equation just written to Equation 8.1, we deduce that:

$$a = 1; \quad b = 4; \quad c = -4$$

Consequently:

$$x = \frac{-4 \pm \sqrt[2]{16 + 16}}{2} = \frac{-4 \pm \sqrt[2]{32}}{2}$$

The square root of 32 is 5.66. So:

$$x = \frac{-4 \pm 5.66}{2} = \frac{1.66}{2} \text{ or } \frac{-9.66}{2}$$

$$x = 0.83 \text{ or } -4.83$$

Quadratic equations, as in the preceding example, always have two roots. When these equations arise in problems in chemistry, it will ordinarily turn out that one of the roots is physically absurd (Example 8.9).

Example 8.9 Consider the equilibrium between a solution of acetic acid and its ions:

$$HC_2H_3O_2(aq) \rightleftharpoons H^+(aq) + C_2H_3O_2^-(aq); \quad K_a = 1.80 \times 10^{-5} .$$

In Example 8.7, we derived the following algebraic equation for the equilibrium concentration of H^+, x, in a 0.10 molar solution of acetic acid.

$$1.80 \times 10^{-5} = \frac{x^2}{0.100 - x}$$

Solve this equation for x, using the quadratic formula.

Solution Rearranging the equation to get it in standard form:

$$x^2 = 1.80 \times 10^{-6} - 1.80 \times 10^{-5}x$$

$$x^2 + 1.80 \times 10^{-5}x - 1.80 \times 10^{-6} = 0$$

$$a = 1; \quad b = 1.80 \times 10^{-5}; \quad c = -1.80 \times 10^{-6}$$

Hence, $x = \dfrac{-1.80 \times 10^{-5} \pm \sqrt[2]{(3.24 \times 10^{-10}) + 7.20 \times 10^{-6}}}{2}$

To obtain the square root, we note that: $3.24 \times 10^{-10} = 0.000324 \times 10^{-6}$. Hence:

$$3.24 \times 10^{-10} + 7.20 \times 10^{-6} = 0.000324 \times 10^{-6} + 7.20 \times 10^{-6}$$

$$\approx 7.20 \times 10^{-6}$$

$$x = \frac{-1.80 \times 10^{-5} \pm \sqrt[2]{7.20 \times 10^{-6}}}{2}$$

$$= \frac{-1.80 \times 10^{-5} \pm 2.68 \times 10^{-3}}{2}$$

$$= \frac{-0.0180 \times 10^{-3} \pm 2.68 \times 10^{-3}}{2}$$

$$= \frac{2.66 \times 10^{-3}}{2} \quad \text{or} \quad \frac{-2.70 \times 10^{-3}}{2}$$

Clearly, the second root is absurd; we cannot have a *negative* concentration of H^+ ions. We deduce that $x = 1.33 \times 10^{-3}$.

EXERCISES

1. Solve the following equations for x:
   a. $3x^2 = 4.5 \times 10^{-9}$
   b. $x^2/(1 - x)^2 = 2.0$
   c. $x^2/(1 - x) = 0.30$
   d. $x/(1 - 2x)^2 = 5.0$
   e. $4x^2/(2 - x) = 20$
   f. $x^4/(2 - x)^2 = 1.0 \times 10^{-4}$

2. Referring to Example 8.9, solve for the concentration of H^+ if:
   a. The original concentration of weak acid is 1.00 instead of 0.100.
   b. K_a is 1.80×10^{-2} instead of 1.80×10^{-5}.

8.5 SOLUTION OF SECOND DEGREE EQUATIONS BY APPROXIMATION METHODS

Although it is always possible to solve a second degree equation with the aid of the quadratic formula, it is seldom convenient to do so. Example 8.9 illustrates the rather complicated and tedious arithmetic that is involved. Since second degree equations similar to that encountered in Example 8.9 arise very frequently in general chemistry, it is highly desirable to find simpler ways of solving them. One such approach, often referred to as the method of *successive approximations*, can be used with a wide variety of problems dealing with equilibria in solutions of weak acids and bases.

To illustrate how this method works, let us apply it to Example 8.9, where we were required to solve the equation:

$$\frac{x^2}{0.100 - x} = 1.80 \times 10^{-5}$$

Noting that the ionization constant of acetic acid, 1.80×10^{-5}, is a very small number, it seems reasonable to suppose that x, the concentration of H^+ produced when acetic acid ionizes, will be very small. In particular, it seems likely that x *will be very much smaller than 0.100*, the original concentration of acetic acid. If this is true, we would be justified in ignoring the x in the denominator of the above equation, writing:

$$\frac{x^2}{0.100} = 1.80 \times 10^{-5}$$

This approximate equation is readily solved for x:

$$x^2 = 1.80 \times 10^{-6}; \quad x = \sqrt{1.80} \times 10^{-3} = 1.34 \times 10^{-3}$$

Let us compare this value of x, obtained by making the approximation $0.100 - x \approx 0.100$, to that obtained in Example 8.9, where we used the quadratic formula. The two numbers, 1.34×10^{-3} and 1.33×10^{-3}, differ from each other by 1 part in 133, or less than 1 per cent. Errors of this order of magnitude are ordinarily acceptable in working problems dealing with weak electrolyte equilibria, since the equilibrium constants themselves are seldom valid to better than ±5 per cent.

Clearly, the simplifying assumption described above is valid in the particular case dealt with in Example 8.9. The question of the general validity of this approach arises. Specifically, if we are dealing with an equation of the type:

$$\frac{x^2}{a - x} = K \tag{8.3}$$

we wish to know under what conditions it is legitimate to ignore the x in the denominator and make the approximation:

$$x^2 \approx aK; \quad x \approx (aK)^{1/2}$$

A general rule that we will use in this situation is to **regard the approximation as valid, provided the value of x obtained is no more than 5% of a.** That is:

$$a - x \approx a \text{ if } x \leqslant 5\% \, a \tag{8.4}$$

In most problems dealing with equilibria of weak acids or bases, we will find, using the preceding criterion, that the approximation is valid. In the case of 0.100 molar acetic acid, for example, the value of x calculated by making the approximation is only a little more than 1 per cent of the original concentration:

$$\frac{x}{a} = \frac{1.34 \times 10^{-3}}{1.00 \times 10^{-1}} = 0.0134 = 1.34\%$$

Occasionally, however, the error arising from this approximation will exceed the 5 per cent limit that we have set. If this should happen, we need not give up. We can always refine our calculation by making a second approximation, more nearly valid than the first. What we do here is to substitute for x, in the denominator of Equation 8.3, the value obtained by the first approximation. Solving the resultant equation for x gives a number considerably closer to the true value. The technique is illustrated in Example 8.10.

Example 8.10 The weak acid HSO_4^- has an ionization constant of 1.0×10^{-2}

$$HSO_4^-(aq) \rightleftharpoons H^+(aq) + SO_4^{2-}(aq); \quad K_a = 1.0 \times 10^{-2}$$

Calculate the equilibrium concentration of H^+ in a solution prepared by adding one mole of HSO_4^- to a liter of water.

Solution Following the same line of reasoning as in Example 8.7, we arrive at the following table.

	Orig. Conc. (Mole/Liter)	Change	Equil. Conc. (Mole/Liter)
HSO_4^-	1.0	$-x$	$1.0 - x$
H^+	0.0	$+x$	x
SO_4^{2-}	0.0	$+x$	x

This leads to the algebraic equation:

$$\frac{x^2}{1.0 - x} = 1.0 \times 10^{-2}$$

Making the approximation: $1.0 - x \approx 1.0$, we obtain:

$$x^2 \approx 1.0 \times 10^{-2}; \quad x \approx 1.0 \times 10^{-1} = 0.10$$

In this case, the error exceeds that allowed under the "5 per cent rule" (Equation 8.4).

$$\frac{x}{a} = \frac{0.10}{1.0} = 0.10 = 10\%$$

To obtain a more accurate value, we substitute $x = 0.10$ in the denominator of the original equation and obtain:

$$\frac{x^2}{1.0 - 0.10} = \frac{x^2}{0.90} = 1.0 \times 10^{-2}$$

$$x^2 = 0.90 \times 10^{-2}; \quad x = 0.95 \times 10^{-1} = 0.095 \text{ mole/liter} = [H^+]$$

This value of $[H^+]$ is closer to the true concentration, since 0.90 is a better approximation to the equilibrium concentration of HSO_4^- than was our first guess, 1.0. If we are still not satisfied, we might attempt a further improvement, using the value of $[H^+]$ just calculated to obtain a still better value for $[HSO_4^-]$. If we make this improvement, we find that $[HSO_4^-]$ stays at 0.90. That is:

$$[HSO_4^-] = 1.0 - 0.095 = 0.90$$

This calculation means that if we were to solve again for $[H^+]$, we would get the same answer.

This technique of successive approximations can be applied to a wide variety of problems dealing with equilibria in solutions of weak acids and bases. Ordinarily, a single approximation will be sufficient to obtain an answer that meets the 5 per cent criterion. Sometimes, as in Example 8.10, a second approximation will be necessary. Almost never do we need to go beyond this point.

The approximation method we have described is by no means restricted to second degree equations. Indeed, it is particularly useful for higher degree equations where exact solutions are either very difficult or impossible to obtain. Consider, for example, the equation derived for Exercise 3 of Section 8.3 where, for the equilibrium:

$$2CO_2(g) \rightleftharpoons 2CO(g) + O_2(g); \quad K_c = 5.0 \times 10^{-4}$$

you should have obtained the following algebraic equation:

$$\frac{4x^3}{(1 - 2x)^2} = 5.0 \times 10^{-4}$$

Noting that the equilibrium constant for this reaction is small, we might assume that:

$$(1 - 2x) \approx 1$$

The equation would then become:

$$4x^3 \approx 5.0 \times 10^{-4}$$

$$x^3 \approx 1.25 \times 10^{-4} = 125 \times 10^{-6}$$

$$x \approx 5.0 \times 10^{-2} = 0.050$$

If we wish to refine our answer, we make a second approximation, substituting this value of x in the denominator of the original equation.

$$\frac{4x^3}{(1 - 0.10)^2} = \frac{4x^3}{(0.90)^2} \approx 5.0 \times 10^{-4}$$

$$4x^3 \approx (0.81)(5.0 \times 10^{-4}) = 4.05 \times 10^{-4}$$

$$x^3 \approx 1.01 \times 10^{-4} = 101 \times 10^{-6}$$

$$x \approx 4.7 \times 10^{-2} = 0.047$$

You can readily demonstrate that if a third approximation is made, the value of x remains unchanged.

Finally, we should point out that certain equations of the second degree or higher can be solved by making a different kind of approximation. Example 8.11 illustrates such a case.

Example 8.11 For the reaction: $H^+(aq) + C_2H_3O_2^-(aq) \rightleftharpoons HC_2H_3O_2(aq)$, $K_c = 5.6 \times 10^4$. If we start with concentrations of H^+ and $C_2H_3O_2^-$ of 1.0 mole/liter, what are the equilibrium concentrations of all species?

Solution If we let x = equilibrium concentration of $HC_2H_3O_2$, the table becomes:

	Orig. Conc. (Mole/Liter)	Change	Equil. Conc. (Mole/Liter)
H^+	1.0	$-x$	$1.0 - x$
$C_2H_3O_2^-$	1.0	$-x$	$1.0 - x$
$HC_2H_3O_2$	0.0	$+x$	x

Consequently, the algebraic equation is:

$$K = \frac{[HC_2H_3O_2]}{[H^+] \times [C_2H_3O_2^-]} = \frac{x}{(1.0 - x)^2} = 5.6 \times 10^4$$

Now, since the equilibrium constant is very large, most of the H^+ and $C_2H_3O_2^-$ must react to form $HC_2H_3O_2$; *this means that (1.0 − x) will be very small or that $x \approx 1$.* If we make this substitution in the *numerator* of the above equation:

$$\frac{1}{(1.0 - x)^2} = 5.6 \times 10^4$$

Solving: $(1 - x)^2 = \dfrac{1}{5.6 \times 10^4} = 1.8 \times 10^{-5} = 18 \times 10^{-6}$

$$1 - x = 4.2 \times 10^{-3} = \text{conc. } C_2H_3O_2^- = \text{conc. } H^+$$

$$\text{conc. } HC_2H_3O_2 = x = 1 - 4.2 \times 10^{-3} \approx 1$$

EXERCISES

1. Solve the following equations by making the approximation: $(a - x) \approx a$.

   a. $\dfrac{x^2}{1 - x} = 1.0 \times 10^{-6}$

   b. $\dfrac{x^2}{0.10 - x} = 1.0 \times 10^{-3}$

   c. $\dfrac{x^2}{0.20 - x} = 2.0 \times 10^{-4}$

2. Following the 5 per cent rule, solve the following equations, using one or at most two approximations.

   a. $\dfrac{x^2}{1.0 - x} = 1.0 \times 10^{-4}$ b. $\dfrac{x^2}{0.020 - x} = 1.0 \times 10^{-3}$

   c. The equation obtained in Exercise 1, Section 8.3.

3. Solve the following equations by making successive approximations until the value of x remains the same to two significant figures.

   a. $\dfrac{x^3}{(2.0 - x)^2 (1.0 + x)} = 2.0 \times 10^{-6}$

   b. $\dfrac{x^2}{1.00 - x} = 50 \quad (0 < x < 1)$

8.6 SIMULTANEOUS FIRST DEGREE EQUATIONS IN TWO UNKNOWNS

Occasionally, problems in general chemistry lead naturally to a system of two equations in two unknowns. All of the equations of this type that we will encounter will be of the first degree in both unknowns. A typical example would be:

$$2x + 5y = 15$$

$$3x - 4y = 1.8$$

A simple way to solve a pair of equations of this type is to use one of the equations to eliminate one variable. In the example just cited, we might solve the first equation for x:

$$2x = 15 - 5y; \quad x = (15 - 5y)/2$$

and substitute in the second equation:

$$3\frac{(15 - 5y)}{2} - 4y = 1.8$$

We now have a first degree equation in one unknown, which is readily solved by the method discussed in Section 8.1 to obtain:

$$y = 1.8$$

Having obtained a numerical value for y, we can easily find x, using either of the two original equations. For example, from the first equation:

$$2x + 5(1.8) = 15; \quad 2x = 6; \quad x = 3$$

To illustrate the use of this technique in a chemical problem, let us return for a moment to Example 8.6, which involved calculating the weight per cent of SnO in a mixture with SnO_2. You will recall that we found that

$$1.000 \text{ g SnO} \rightarrow 0.881 \text{ g Sn}$$

$$1.000 \text{ g SnO}_2 \rightarrow 0.778 \text{ g Sn}$$

Now, suppose we had let x = no. of grams of SnO and y = no. of grams of SnO_2. Then, noting that the 1.000 gram sample yielded 0.850 grams of tin, we could have written down the two equations:

$$x + y = 1.000$$

$$x(0.881) + y(0.778) = 0.850$$

Solving the first equation for y and substituting in the second:

$$x(0.881) + (1.000 - x)0.778 = 0.850$$

You will notice that this is precisely the equation that we arrived at in Example 8.6 by a line of reasoning that avoided introducing the second unknown. The problem from this point on is identical with the one already worked. The answers are:

$$x = 0.70; \quad y = 0.30$$

and, as before, we find the percentages of SnO and SnO_2 to be 70 and 30, respectively.

EXERCISES

1. Solve for x and y:
   a. $3x - 9y = 16$ b. $x + y = 0.200$
      $2x + 5y = -9$ $0.0556x + 0.0153y = 0.00700$

2. Solve for x, y, and z:

   $2x + 3y - z = 6$

   $x - 4y + 2z = 9$

   $3x + y - 3z = 4$

(It is always possible to solve independent simultaneous equations to obtain numerical answers for each unknown, provided the number of equations equals the number of unknowns. First degree equations of this type can be solved by successively eliminating unknowns.)

PROBLEMS

8.1 The relationship between temperatures expressed in $°K$ and in $°C$ is:

$$°K = °C + 273°$$

Combine this equation with that given in Example 8.1 to obtain a relationship between $°F$ and $°K$.

8.2 Using the ideal gas law, $PV = nRT$, calculate the temperature at which a sample of N_2 weighing 20.0 g will occupy a volume of 6.20 liters at a pressure of 3.15 atm ($R = 0.0821$ liter atm/mole $°K$).

8.3 The minimum energy in ergs, ϵ, of a particle of mass m (grams) confined to move in a small box of volume V (cm^3) is

$$\epsilon = \frac{h^2}{8m V^{2/3}}$$

where h = Planck's constant = 6.62×10^{-27} erg sec.

a. What must be the mass of a particle if it is to have a minimum energy of 1.0×10^{-11} ergs when confined to a cubical box 1.0×10^{-8} cm on a side?
b. What must be the volume of a box in which a proton ($m = 1.0/6.0 \times 10^{23}$ g) has a minimum energy of 1.0×10^{-6} ergs?

8.4 Calculate the molality (moles solute per kg solvent) of a solution prepared by dissolving 50.0 grams of KCl:
  a. in 520 grams of water.
  b. to form 920 grams of solution.
  c. to form 920 ml of solution (density = 1.033 g/ml).

8.5 Starting with the ideal gas law, $PV = nRT$, derive an equation for the density of an ideal gas ($d = g/V$), and use it to calculate the density of N_2 at $300°K$ and 3.50 atm pressure ($R = 0.0821$ liter atm/mole $°K$).

8.6 The Van der Waals equation for one mole of a non-ideal gas is:

$$\left(P + \frac{a}{V^2}\right)(V - b) = RT$$

where a and b are constants for a particular gas and the other symbols have the same meaning as in the ideal gas law. For N_2, $a = 1.39$ atm liter^2, $b = 0.0391$ liter. Using this equation, calculate the pressure exerted by one mole of N_2 in a 1.00 liter container at $300°K$, and compare to the value calculated using the ideal gas law.

8.7 The molarity, c, of a solution is related to the molality, m, by the expression:

$$c = \frac{md}{1 + mM_2/1000}$$

where d is the density in g/ml and M_2 is the molar weight of solute.
  a. Solve this equation for m.
  b. Calculate the molality of a 0.1000 molar solution which has a density of 1.004 g/ml and contains a solute of molar weight 184.

8.8 Analysis of a 2.000 gram sample consisting of a mixture of LiCl and KCl shows that it contains 1.200 grams of chlorine. What are the percentages of LiCl and KCl in the mixture?

8.9 Set up the equilibrium constant expressions for each of the following.
  Example: $PCl_5(g) \rightleftharpoons PCl_3(g) + Cl_2(g)$; $K_c = 5.0$

$$5.0 = \frac{[PCl_3] \times [Cl_2]}{[PCl_5]}$$

  a. $Cu(NH_3)_4^{2+}(aq) \rightleftharpoons Cu^{2+}(aq) + 4NH_3(aq)$; $K_c = 2 \times 10^{-13}$
  b. $2H_2O(g) \rightleftharpoons 2H_2(g) + O_2(g)$; $K_c = 1.0 \times 10^{-9}$
  c. $2NH_3(g) \rightleftharpoons N_2(g) + 3H_2(g)$; $K_c = 2.0 \times 10^4$

8.10 In Problem 8.9, assume that we start in each case with pure reactant (left side of equation) at a concentration of 1.00 mole/liter. Set up the appropriate algebraic equation for each equilibrium, expressing the equilibrium concentrations of all species in terms of a single unknown.

Example: for the $PCl_5 - PCl_3 - Cl_2$ equilibrium: $5.0 = \dfrac{x^2}{1-x}$

8.11 Using the quadratic formula, solve the algebraic equation given in 8.10 for the $PCl_5 - PCl_3 - Cl_2$ case to find the equilibrium concentrations of all species.

8.12 For the dissociation of nitrous acid: $HNO_2(aq) \rightleftharpoons H^+(aq) + NO_2^-(aq)$, $K_a = 4.5 \times 10^{-4}$. Following the 5 per cent rule, estimate the equilibrium concentration of H^+, starting with HNO_2 concentrations of:
    a. 1.0 mole/liter   b. 0.10 mole/liter   c. 0.010 mole/liter

8.13 For the reaction:

$$2SO_2(g) + O_2(g) \rightleftharpoons 2SO_3(g), K_c = 1.0 \times 10^{-4}$$

Using a suitable approximation, calculate the equilibrium concentration of SO_3, starting with conc. SO_2 = conc. O_2 = 1.5 mole/liter.

8.14 For the reaction:

$$H^+(aq) + NH_3(aq) \rightleftharpoons NH_4^+(aq), K_c = 1.8 \times 10^9$$

Using a suitable approximation, calculate the equilibrium concentrations of all species starting with conc. H^+ = conc. NH_3 = 1.0 mole/liter.

FUNCTIONAL RELATIONSHIPS

A variable (y) is said to be a function of another variable (x) if, for various values of x, it is possible to establish corresponding values of y. In mathematical symbolism, we describe this situation by writing:

$$y = f(x)$$

The variable (x) to which we first assign numerical values is referred to as the **independent variable**; the other variable (y) is called the **dependent variable**.

The functional relationship between two variables may be expressed in any of three different ways.

1. An *algebraic equation* which makes it possible to calculate the value of y corresponding to any value of x. A simple example is:

$$y = 2x$$

which tells us that y can always be found by multiplying x by 2.

2. A *table* which lists values of y corresponding to selected values of x. Table 9.1 does this for the function $y = 2x$ at six different values of x.

3. A *graph* drawn through points indicating corresponding values of y and x. The graph of the function $y = 2x$ between $x = 0$ and $x = 5$ is shown in Figure 9.1; the points are those listed in Table 9.1.

In this chapter, we will consider the several types of functional relationships that arise in general chemistry and see how they can be expressed by means of equations and tables. A separate chapter (Chapter 10) will be devoted to ways of representing functional relationships by means of graphs.

Table 9.1

y	0	2	4	6	8	10
x	0	1	2	3	4	5

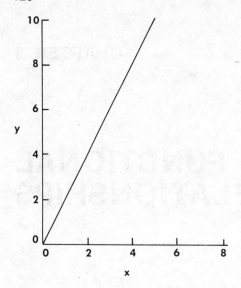

FIGURE 9.1   Graph of $y = 2x$.

9.1   DIRECT PROPORTIONALITY

$$y = ax; \quad y_2/y_1 = x_2/x_1$$

Many functional relationships in general chemistry fall in this category. Listed in Table 9.2 are two examples in which one variable (volume of a gas sample at constant pressure, rate of decomposition of N_2O_5) is **directly proportional** to another (absolute temperature, concentration of N_2O_5). Looking at these two sets of data, we see that they have a common characteristic. In both cases, *the quotient y/x has a constant value throughout the set* (0.082 in the first case, 0.030 in the second).

Algebraically, we can say that y is directly proportional to x if, for all values of the two variables:

$$\frac{y}{x} = a; \quad \text{or,} \quad y = ax \tag{9.1}$$

Table 9.2   **Examples of Direct Proportionalities in General Chemistry**

1. Volume, V, of one mole of an ideal gas at 1 atm as a function of absolute temperature, T.

V (liters)	8.2	16.4	24.6	32.8	41.0
T ($^\circ K$)	100	200	300	400	500
V/T	0.082	0.082	0.082	0.082	0.082

2. Rate of decomposition of N_2O_5 at 45°C as a function of concentration of N_2O_5.

Rate (mole/liter min)	0.0030	0.0060	0.0090	0.0120	0.0150
Conc. (mole/liter)	0.10	0.20	0.30	0.40	0.50
Rate/Conc.	0.030	0.030	0.030	0.030	0.030

where a is a constant, having the same value throughout. Clearly, the two functional relationships listed in Table 9.2 meet this criterion, as does the relationship in Table 9.1 ($y = 2x$).

The defining equation for direct proportionality, Equation 9.1, leads to a simple "two-point" equation which can be used to relate "final" and "initial" values of y to the corresponding values of x. Using the subscripts 2 and 1 to represent final and initial states, respectively, we have:

$$y_2 = ax_2$$

$$y_1 = ax_1$$

Dividing the first equation by the second, the constant a drops out, and we have:

$$y_2/y_1 = x_2/x_1; \quad \text{or} \quad y_2 = y_1 x_2/x_1 \tag{9.2}$$

Example 9.1 The solubility of a gas in a liquid is directly proportional to its partial pressure. The solubility of O_2 in water at $25°C$ is 0.085 mole/liter at one atmosphere. Calculate its solubility at a partial pressure of 0.20 atm.

Solution Using the letters S and P to represent solubility and pressure, respectively, we have, from Equation 9.2:

$$S_2 = S_1 \times \frac{P_2}{P_1}$$

Taking P_1 to be one atmosphere and P_2 to be 0.20 atmosphere, the solubility, S_2, at the latter pressure is:

$$S_2 = 0.085 \frac{mole}{liter} \times \frac{0.20 \ atm}{1.00 \ atm} = 0.017 \frac{mole}{liter}$$

An alternative approach would be to use the solubility at one atmosphere to calculate the proportionality constant.

$$a = \frac{S_1}{P_1} = \frac{0.085 \ mole/liter}{1 \ atm} = 0.085 \ mole/liter \ atm$$

Then, knowing the value of a, use it to obtain S_2:

$$S_2 = aP_2 = 0.085 \frac{mole}{liter \ atm} \times 0.20 \ atm = 0.017 \ mole/liter$$

Sometimes in chemical relationships we find that y is directly proportional not to x itself, but to some positive power of x. Thus we find that the average velocity,

u, of a gas molecule is directly proportional to the square root of the absolute temperature, T.

$$u = aT^{1/2}$$

In another case, the rate of decomposition of acetaldehyde, CH_3CHO, is directly proportional to the square of its concentration.

$$rate = k(conc. \ CH_3CHO)^2$$

Relations entirely analogous to Equation 9.2 can be derived for cases such as these (Example 9.2).

Example 9.2 The rate of decomposition of acetaldehyde is 0.19 mole/liter min when its concentration is 0.010 mole/liter. What is the rate of decomposition when the concentration increases to 0.020 mole/liter?

Solution Writing the rate equation twice, for final and initial states, we have:

$$rate_2 = k(conc._2)^2$$
$$rate_1 = k(conc._1)^2$$

Dividing:

$$\frac{rate_2}{rate_1} = \frac{(conc._2)^2}{(conc._1)^2}; \quad rate_2 = rate_1 \times \frac{(conc._2)^2}{(conc._1)^2} = rate_1 \times \left(\frac{conc._2}{conc._1}\right)^2$$

Substituting numbers:

$$rate_2 = 0.19 \ \frac{mole}{liter \ min} \times \left(\frac{0.020 \ mole/liter}{0.010 \ mole/liter}\right)^2 = 0.19 \ \frac{mole}{liter \ min} \times 4$$

$$= 0.76 \ \frac{mole}{liter \ min}$$

EXERCISES

1. Given that y is directly proportional to x, calculate the value of y when $x = 7$, if, when $x = 2$, y is:
   a. 1   b. 2   c. 7   d. −5   e. 0.5

2. For each part of (1), calculate the value of the proportionality constant.

3. Given that $y = ax$, write an equation for:
   a. $y_2 + y_1$ b. $y_2 - y_1$ c. $(y_2 - y_1)/(x_2 - x_1)$

4. Which of the following functions are of the type $y = ax$? $y = ax^2$? $y = ax^{1/2}$?
   Where possible, fill in the missing numbers.

a. y	x	b. y	x	c. y	x	d. y	x	e. y	x
0	0	2	1	0	0	0	0	1	0
5	1	8	2	3	1	2	1	3	1
10	2	18	3	6	2	2.8	2	5	2
15	3	—	4	9	3	3.5	3	7	3
—	4			11	—	—	4	—	4

9.2 INVERSE PROPORTIONALITY

$$y = a/x; \qquad y_2/y_1 = x_1/x_2$$

A quantity y is said to be **inversely proportional** to x if their product is the same for all values of x, i.e.:

$$yx = a; \qquad \text{or} \qquad y = a/x \tag{9.3}$$

where a is a constant. Two familiar examples of inverse proportionalities in general chemistry are described in Table 9.3. Note the constancy of the product of the two variables throughout the range of values given.

Table 9.3 Examples of Inverse Proportionality in General Chemistry

1. Volume, V, of one mole of an ideal gas at $25°C$ as a function of pressure, P.

V (liters)	24.5	12.2	8.16	6.12	4.89
P (atm)	1.00	2.00	3.00	4.00	5.00
$P \times V$	24.5	24.4	24.5	24.5	24.4

2. Conc. H^+ in aqueous solution at $25°C$ as a function of conc. OH^-.

Conc H^+ (mole/liter)	1.0×10^{-14}	1.0×10^{-10}	1.0×10^{-7}	1.0×10^{-4}
Conc OH^- (mole/liter)	1.0	1.0×10^{-4}	1.0×10^{-7}	1.0×10^{-10}
Conc $H^+ \times$ conc OH^-	1.0×10^{-14}	1.0×10^{-14}	1.0×10^{-14}	1.0×10^{-14}

Example 9.3 Making use of the information in Table 9.3, calculate the concentration of OH^- in a solution in which conc. $H^+ = 5.0 \times 10^{-2}$.

Solution The fundamental relation is:

$$\text{conc. } H^+ \times \text{conc. } OH^- = 1.0 \times 10^{-14}$$

Solving for conc. H^+:

$$\text{conc. } H^+ = \frac{1.0 \times 10^{-14}}{\text{conc. } OH^-} = \frac{1.0 \times 10^{-14}}{5.0 \times 10^{-2}} = 0.20 \times 10^{-12}$$

$$= 2.0 \times 10^{-13} \text{ mole/liter}$$

The two-point equation corresponding to an inverse proportionality is readily obtained by writing Equation 9.3 for both final and initial states.

$$y_2 x_2 = a; \quad y_1 x_1 = a$$

So:

$$y_2 x_2 = y_1 x_1$$

or:

$$\frac{y_2}{y_1} = \frac{x_1}{x_2}; \quad y_2 = y_1 \frac{x_1}{x_2} \tag{9.4}$$

One of the fundamental laws concerning the physical behavior of gases, Graham's Law of Effusion, is ordinarily stated in terms of an inverse proportionality. We find that the rate of effusion of a gas at a given temperature and pressure is inversely proportional to the square root of its molecular weight, M:

$$\text{rate} \times M^{1/2} = \text{constant}$$

or:

$$\frac{\text{rate}_2}{\text{rate}_1} = \frac{(M_1)^{1/2}}{(M_2)^{1/2}} = \left(\frac{M_1}{M_2}\right)^{1/2}$$

where the subscripts 2 and 1 refer to two gases of different molecular weight.

Example 9.4 It is found experimentally that oxygen, O_2, effuses 2.41 times as fast as a certain gas of unknown molecular weight. What is the molecular weight of the gas?

Solution Using the subscript 2 to represent O_2 and 1 for the other gas, we have:

$$\frac{\text{rate}_2}{\text{rate}_1} = 2.41 = \left(\frac{M_1}{32.0}\right)^{1/2}$$

To solve, we square both sides of the equation.

$$(2.41)^2 = \frac{M_1}{32.0}$$

$$M_1 = 32.0(2.41)^2 = 32.0 \times 5.81 = 186$$

EXERCISES

1. Assuming that y is inversely proportional to x, complete the following tables.

   a. y ___ 4.0 6.0   b. y ___ ___ 3.0 ___
      x 2.0 3.0 ___ x 0.010 0.10 1.0 10

2. In which of the following is y inversely proportional to x? directly proportional to x? neither?

a.	y	2.0	1.0	b.	y	1.0	2.0	c.	y	4	8	-2	d.	y	3.0
	x	1.0	2.0		x	1.0	2.0		x	-2	-1	4		x	0.0

3. Assuming y is inversely proportional to x, what is the value of y_2/y_1 when:
   a. $x_1/x_2 = 5$ b. $x_2/x_1 = 5$ c. $(x_2 - x_1)/x_2 = 5$

4. If $yx^2 = a$, derive an expression for y_2/y_1.

9.3 LINEAR FUNCTIONS

$$y = ax + b; \quad (y_2 - y_1)/(x_2 - x_1) = a$$

A **linear function** is one which has the general form:

$$y = ax + b \tag{9.5}$$

where a and b are constants. The phrase "linear function" is used because a straight line is obtained when y is plotted against x (Chapter 10). The relationship: $y = ax$, discussed in Section 9.1, is a special case of a linear function with $b = 0$.

The most useful two-point equation for a linear function is obtained by writing Equation 9.5 for the two states:

$$y_2 = ax_2 + b$$

$$y_1 = ax_1 + b$$

and subtracting:

$$y_2 - y_1 = a(x_2 - x_1); \quad \text{or} \quad \frac{y_2 - y_1}{x_2 - x_1} = a \tag{9.6}$$

Equation 9.6 offers perhaps the simplest way to check a table of data to see if it represents a linear function (Table 9.4); it is also useful in many problems in chemistry (Example 9.5).

Table 9.4   Test for a Linear Function: $(y_2 - y_1)/(x_2 - x_1)$ = constant

Function 1			Function 2			Function 3		
y	x	$\dfrac{(y_2 - y_1)}{(x_2 - x_1)}$	y	x	$\dfrac{(y_2 - y_1)}{(x_2 - x_1)}$	y	x	$\dfrac{(y_2 - y_1)}{(x_2 - x_1)}$
3	0		1	0		0	0	
5	1	2/1 = 2	4	1	3/1 = 3	3	1	3/1 = 3
7	2	2/1 = 2	9	2	5/1 = 5	6	2	3/1 = 3
9	3	2/1 = 2	16	3	7/1 = 7	9	3	3/1 = 3
linear			*non-linear*			*linear*		

Example 9.5 Temperatures expressed in °F are a linear function of temperatures expressed in °C. The relation is:

$$°F = 1.8°C + 32°$$

a. If the centigrade temperature rises by 15°, what is the corresponding increase in °F?
b. A 20° increase in °F corresponds to what change in °C?

Solution

a. In this equation, °F represents "y", °C is "x", $a = 1.8$, $b = 32°$. From Equation 9.6:

$$\frac{F_2 - F_1}{C_2 - C_1} = 1.8$$

We are given that the increase in °C is 15°; hence the increase in °F must be:

$$1.8(15°) = 27°$$

b. Here, the increase in °F is 20°; hence:

$$\text{increase in } °C = \frac{\text{increase in } °F}{1.8} = \frac{20}{1.8} = 11°$$

One of the fundamental equations of thermodynamics, the so-called Gibbs-Helmholtz equation, can be considered as a linear function relating the free energy change, ΔG, of a reaction to the absolute temperature, T:

$$\Delta G = \Delta H - T\Delta S$$

The quantities ΔH and ΔS appearing in this equation are "constants" in the sense that they do not vary, at least to a first approximation, with temperature. These quantities are referred to as the enthalpy change and entropy change of a reaction, respectively.

Example 9.6 Using the previously mentioned equation, determine:
a. ΔG at 400°K for a reaction for which $\Delta H = 61.0$ kcal and $\Delta S = 0.020$ kcal/°K.
b. ΔG at 500°K for a reaction for which $\Delta H = -32.0$ kcal and $\Delta G = -20.0$ kcal at 300°K.

Solution

a. $\Delta G = 61.0$ kcal $- 400°K(0.020$ kcal/°K$) = 53.0$ kcal.
b. One way to solve this problem is to first calculate ΔS by applying the equation at 300°K, and then use this value to calculate ΔG at 500°K.

At 300°K: -20.0 kcal $= -32.0$ kcal $- 300°K(\Delta S)$
$\Delta S = -12.0$ kcal/300°K $= -0.040$ kcal/°K

At 500°K: $\Delta G = -32.0$ kcal $- 500°K (-0.040$ kcal/°K$)$
$= -12.0$ kcal

EXERCISES

1. Which of the following are linear functions? direct proportionalities? neither?
   a. y 3.6 6.0 8.4   b. y 2.2 3.4 4.6
      x 3.0 5.0 7.0 x 1.0 2.0 3.0
   c. y 0.5 3.0 8.0   d. y 5.0 2.5 1.7
      x 1.0 2.0 4.0 x 1.0 2.0 3.0

2. For each of the linear functions in (1), give the values of a and b.

3. For the linear function, $y = 3x - 4$, give:
   a. $(y_2 - y_1)/(x_2 - x_1)$ b. $(y_2 - y_1)$ when $(x_2 - x_1) = -6.0$
   c. $(x_2 - x_1)$ when $(y_2 - y_1) = -4.0$

9.4 LOGARITHMIC FUNCTIONS

$$\log y = \frac{-A}{x} + B; \quad \log \frac{y_2}{y_1} = \frac{A(x_2 - x_1)}{x_2 x_1}$$

In many of the functional relationships that we work with in general chemistry, one of the variables appears as a logarithmic term (Chapter 5). Perhaps the most common logarithmic function in chemistry takes the form:

$$\log y = \frac{-A}{x} + B \tag{9.7}$$

where A and B are positive numbers. Three examples of functional relationships of this type are given in Table 9.5. Note that in each case x is the temperature in °K, while A includes an energy term (heat of vaporization, heat of reaction, energy of activation). The constant B may be thought of as the limiting value approached by $\log y$ at very high temperatures $(-A/x \rightarrow 0$ as $x \rightarrow \infty)$.

Table 9.5 Examples of the Function: $\log y = \dfrac{-A}{x} + B$

Function	A	Meaning of Terms
1. $\log P = \dfrac{-\Delta H}{2.30\,RT} + B$	$\dfrac{\Delta H}{2.30\,R}$	P = vapor pressure of liquid T = absolute temperature (°K) R = gas constant = 1.99 cal/mole °K ΔH = heat of vaporization in cal/mole
2. $\log K = \dfrac{-\Delta H}{2.30\,RT} + B$	$\dfrac{\Delta H}{2.30\,R}$	K = equilibrium constant for reaction at absolute temperature T. ΔH = enthalpy change in reaction (cal)
3. $\log k = \dfrac{-\Delta E_a}{2.30\,RT} + B$	$\dfrac{\Delta E_a}{2.30\,R}$	k = rate constant at T ΔE_a = activation energy (cal)

Example 9.7 The vapor pressure, in atmospheres, of benzene is given as a function of temperature by the equation:

$$\log P \text{ (atm)} = \frac{-1785}{T} + 5.08$$

   a. Evaluate the vapor pressure of benzene at 25°C.
   b. Estimate the heat of vaporization of benzene, using the relation given in Table 9.5.
   c. How should the above equation be rewritten to give directly the vapor pressure of benzene in mm Hg?

Solution

   a. 25°C = 298°K. Substituting:

$$\log P \text{ (atm)} = \frac{-1785}{298} + 5.08 = -5.99 + 5.08 = -0.91 = 0.09 - 1$$

The antilog of 0.09 is about 1.2 (see Chapter 5, Logarithms). Hence:

$$P = 1.2 \times 10^{-1} \text{ atm} = 0.12 \text{ atm}$$

   b. From Table 9.5, we see that:

$$A = \Delta H_{vap}/2.30 \, R.$$

Hence:

$$\Delta H_{vap} = 2.30 \, RA = (2.30)(1.99)(1785) \text{ cal} = 8170 \text{ cal}$$

   c. The pressure in mm Hg could be found by multiplying the pressure in atm by 760 (1 atm = 760 mm Hg). Thus:

$$\log P \text{ (mm Hg)} = \log P \text{ (atm)} + \log 760$$

$$= \log P \text{ (atm)} + 2.88$$

Substituting for $\log P$ (atm) in the original equation:

$$\log P \text{ (mm Hg)} - 2.88 = \frac{-1785}{T} + 5.08$$

$$\log P \text{ (mm Hg)} = \frac{-1785}{T} + 7.96$$

Note that what we have done, in effect, is to increase the constant B by 2.88 units (log 760 = 2.88).

The two-point equation corresponding to Equation 9.7 is obtained in a manner similar to that used for a linear function.

$$\log y_2 = \frac{-A}{x_2} + B$$

$$\log y_1 = \frac{-A}{x_1} + B$$

Subtracting:

$$\log y_2 - \log y_1 = \frac{-A}{x_2} + \frac{A}{x_1}$$

$$\log \frac{y_2}{y_1} = A\left[\frac{1}{x_1} - \frac{1}{x_2}\right]$$

This equation is ordinarily written in a somewhat different form, obtained by putting the quantity in brackets over a common denominator, $x_2 x_1$.

$$\frac{1}{x_1} - \frac{1}{x_2} = \frac{x_2}{x_1 x_2} - \frac{x_1}{x_1 x_2} = \frac{x_2 - x_1}{x_1 x_2}$$

Hence:

$$\log \frac{y_2}{y_1} = A\left[\frac{x_2 - x_1}{x_1 x_2}\right] \tag{9.8}$$

(Note that the constant B drops out in this process, as it did in obtaining the two-point equation for a linear function.) Examples of equations in general, chemistry having the form of Equation 9.8 were presented in Table 5.3; calculations involving these equations were illustrated in Problems 5.6 to 5.10, Chapter 5.

EXERCISES

1. Given that: $\log y = \frac{-A}{x} + B$, and the following values of y and x, evaluate A and B.

   a. y 0.100 0.316   b. y 10.0   100
    x 1.00   2.00 x 1.00   2.00

2. In Exercise 1, calculate y when $x = 3.00$

3. Taking $\log y = \frac{-2.00}{x} + 2.00$, complete the following table.

   y ____ ____ 3.00
   x 1.00 2.00 ____

4. Derive the two-point equation corresponding to:
   a. $\log y = -ax + b$ b. $y = -a \log x$

9.5 FUNCTIONS OF MORE THAN ONE VARIABLE

In all the functional relationships discussed to this point (and most of those you will encounter in general chemistry), the dependent variable, y, is expressed as a function of only one independent variable, x. This method of expression reflects the way in which experiments are commonly carried out in the laboratory. If we have reason to believe that a quantity y depends upon more than one variable (i.e., u and z as well as x), we ordinarily design an experiment so that only one of these, x, varies, being careful to hold the others (i.e., u and z) constant. Such an experiment enables us to determine the form of the functional relationship between y and x. Analogous experiments can be carried out to find how y varies with z at constant u and x, with u at constant x and z, and so on. Eventually, we may be able to combine the results of such experiments to obtain a single equation giving the functional dependence of y upon all the independent variables.

This process can be illustrated by the steps involved in finding an expression for the volume, V, of an ideal gas. We know from experience that V varies with three quantities, the pressure, P, the absolute temperature, T, and the number of moles, n. In three separate experiments, we can establish that:

1. At constant n and T, V is inversely proportional to P. That is:

$$V = k_1/P \text{ (constant } n \text{ and } T)$$

2. At constant n and P, V is directly proportional to T.

$$V = k_2 T \text{ (constant, } n, P)$$

3. At constant P and T, V is directly proportional to n.

$$V = k_3 n \text{ (constant } P, T)$$

From the results of these three experiments, it is possible to write a single equation which gives the functional dependence of V upon all three independent variables, P, T, and n. This equation is:

$$V = \frac{\text{constant} \times n \times T}{P}$$

or, in the form in which it is most usually written, the ideal gas law:

$$V = \frac{nRT}{P}; \quad PV = nRT$$

where R is the gas constant. The ideal gas law includes as special cases the three equations that preceded it. Quite clearly, it is more general and hence more useful than any of the three simple relationships.

Another important example in chemistry of a function of more than one variable is the rate of a reaction involving more than one reactant. The rate at which such reactions take place is ordinarily a function of the concentration of more than one species. Consider, for example, the reaction:

$$CO(g) + NO_2(g) \rightarrow CO_2(g) + NO(g)$$

for which the following data can be obtained at $400°C$.

Experiment 1			Experiment 2		
conc. CO	conc. NO_2	rate	conc. CO	conc. NO_2	rate
0.10	0.10	0.005	0.10	0.10	0.005
0.20	0.10	0.010	0.10	0.20	0.010
0.30	0.10	0.015	0.10	0.30	0.015

(Note: Conc. is in mole/liter, rate in mole/liter min.)

From the data of Experiment 1, we see that the rate of reaction is directly proportional to the concentration of CO, holding that of NO_2 constant. Note, for example, that the ratio: rate/conc. CO remains constant at 0.05 throughout Experiment 1. Experiment 2 demonstrates that the rate of reaction is also directly proportional to the concentration of NO_2 (conc. CO constant). A general equation combining these two observations is:

$$rate = k \text{ (conc. CO)} \times \text{(conc. } NO_2)$$

where k is the so-called rate constant for the reaction.

Example 9.8 Given the following data for the reaction between NO and O_2, write the rate expression, which is known to be of the form:

$$rate = k(\text{conc. } O_2)^m (\text{conc. NO})^n$$

where m and n are positive integers (i.e., $1, 2, 3 \cdots$).

conc. NO (mole/liter)	0.10	0.10	0.20
conc. O_2 (mole/liter)	0.10	0.20	0.20
rate (mole/liter min)	0.080	0.16	0.64

Solution The first two columns, where conc. NO remains constant, enable us to determine the functional dependence of the rate of reaction upon the concentration of O_2. It may be obvious to you that a direct proportionality is involved and hence that m, the power to which conc. O_2 is raised in the rate expression, is 1. If not, a general approach is:

$$\frac{rate_2}{rate_1} = \left(\frac{\text{conc.}_2}{\text{conc.}_1}\right)^m$$

Substituting numbers:

$$\frac{0.16}{0.080} = \left(\frac{0.20}{0.10}\right)^m$$

$$2 = 2^m \, ; \, m = 1$$

The second and third entries taken in combination allow us to determine the dependence of rate upon conc. NO, since conc. O_2 is held constant at 0.20. Using the general approach:

$$\frac{rate_3}{rate_2} = \left(\frac{conc._3}{conc._2}\right)^n$$

$$\frac{0.64}{0.16} = \left(\frac{0.20}{0.10}\right)^n$$

$$4 = 2^n \, ; \, n = 2$$

Although it would be desirable to carry out more experiments to definitely establish the rate expression, it would appear from this data to be:

$$rate = k(conc. \, O_2)(conc. \, NO)^2$$

In words, the rate is directly proportional to the concentration of O_2 and to the square of the concentration of NO.

EXERCISES

1. Given the following functional relationships of y with x and z, write a general equation relating y to x and z.

    a. y is directly proportional to x;   directly proportional to z
    b. y is directly proportional to x;   inversely proportional to z
    c. y is directly proportional to x^2;   directly proportional to $z^{1/2}$
    d. y is inversely proportional to x;   inversely proportional to z^2

2. Given the equation: $y = \dfrac{auv^{1/2}}{z}$; $(a = \text{constant})$

    a. Express in words the functional dependence of y upon u, v, and z.
    b. Write a two-point equation relating all variables.
    c. Write a two-point equation relating y to u and z (constant v).

3. Given that $y = k(x)^m (z)^n$, where m and n may be $-1, 0,$ or $+1$, determine m and n.

    a.

y	3	6	3
x	1	2	1
z	1	1	2

    b.

y	-2	-4	-4
x	1	2	1
z	1	1	2

c. y 4 2 8   d. y 3 1 3
 x 1 2 1 x 1 3 1
 z 1 1 2 z 1 1 3

9.6 "CONSTANTS" INVOLVED IN FUNCTIONAL RELATIONSHIPS

Throughout this chapter and indeed throughout the general chemistry course, many different kinds of "constants" are used in expressing functional relationships. Thus we have the gas constant, R, which appears in the ideal gas law, $PV = nRT$, and in a number of thermodynamic relationships (e.g., $\Delta G^\circ = -RT \ln K$). Again, we talk about the rate constant, k, in rate expressions such as:

$$\text{rate} = k(\text{conc. CO})(\text{conc. NO}_2)$$

and the dissociation constant, K_a, in the expression written to describe the equilibrium between acetic acid and its ions in aqueous solution.

$$HC_2H_3O_2(aq) \rightleftharpoons H^+(aq) + C_2H_3O_2^-(aq)$$

$$K_a = \frac{[H^+] \times [C_2H_3O_2^-]}{[HC_2H_3O_2]}$$

In order to use expressions such as these properly, it is important to understand how constants such as R, k, and K_a are determined and what they mean. The following general principles may be helpful in this regard.

1. **Given an equation involving a single constant, the numerical magnitude of the constant can be determined by measuring simultaneously the values of all the variables in the equation.**

Knowing, for example, that one mole of an ideal gas occupies 22.4 liters at $0^\circ C$ ($273^\circ K$) and one atmosphere pressure, we can calculate the gas law constant, R.

$$R = \frac{PV}{RT} = \frac{(1.00 \text{ atm})(22.4 \text{ liters})}{(1.00 \text{ mole})(273^\circ K)} = 0.0821 \frac{\text{liter atm}}{\text{mole }^\circ K}$$

Example 9.9 Given the information in Example 9.8, determine the magnitude of the rate constant k for the reaction of NO with O_2.

Solution You will recall that we established the rate equation to be:

$$\text{rate} = k(\text{conc. O}_2)(\text{conc. NO})^2$$

Substituting the values in the first vertical column of data:

$$0.080 \text{ mole/liter min} = k(0.10 \text{ mole/liter})(0.10 \text{ mole/liter})^2$$

$$k = \frac{0.080}{(0.10)(0.10)^2} = 80 \text{ mole}^{-2} \text{liter}^2 \text{ min}^{-1}$$

If the equation that we are working with involves two constants, two sets of values of the variables will be required to evaluate the constants. For example, given the linear function:

$$y = ax + b$$

we must know two different sets of y and x values to obtain both a and b. Thus, from Equation 9.6, we have:

$$a = \frac{y_2 - y_1}{x_2 - x_1}$$

Knowing y_2, x_2 and y_1, x_1, we can obtain a. Once a is established, b can be calculated by substituting into the original equation at either the final or initial conditions.

$$b = y_2 - ax_2 = y_1 - ax_1$$

In general, *if we have n different constants (i.e., n = 1, 2, 3 · · ·) in an equation, we need n sets of variables to determine the numerical magnitude of each constant.* *

2. **The term "constant" implies that the quantity has the same numerical magnitude regardless of the values taken on by the variables in the equation.**

To illustrate this point, consider the following data for the dissociation of acetic acid in solutions of three different concentrations.

	$[HC_2H_3O_2]$	$[H^+]$	$[C_2H_3O_2^-]$	$K_a = \dfrac{[H^+] \times [C_2H_3O_2^-]}{[HC_2H_3O_2]}$
I	1.0	1.8×10^{-4}	1.0×10^{-1}	$\dfrac{(1.8 \times 10^{-4})(1.0 \times 10^{-1})}{1.0} = 1.8 \times 10^{-5}$
II	1.0×10^{-1}	9.0×10^{-3}	2.0×10^{-4}	$\dfrac{(9.0 \times 10^{-3})(2.0 \times 10^{-4})}{1.0 \times 10^{-1}} = 1.8 \times 10^{-5}$
III	2.0	6.0×10^{-3}	6.0×10^{-3}	$\dfrac{(6.0 \times 10^{-3})(6.0 \times 10^{-3})}{2.0} = 1.8 \times 10^{-5}$

Clearly, K_a remains the same regardless of the values taken on by $[HC_2H_3O_2]$, $[H^+]$, and $[C_2H_3O_2^-]$. This is, of course, a consequence of the fact that there is a functional relationship between these three variables. Only two of them can be chosen independently; once we have made this choice, the other variable is fixed by the requirement that the quotient:

$$[H^+] \times [C_2H_3O_2^-]/[HC_2H_3O_2]$$

always be equal to 1.8×10^{-5}.

---

*Ordinarily, we have more sets of data points than we have constants. For example, we might obtain five sets of data points to evaluate a single constant. The problem then becomes one of determining the best value of the constant(s). Some simple ways of doing this are discussed in Chapter 10 (see also Exercise 2).

3. **The magnitude of a constant may and often does depend upon other variables which are not specified in the equation.**

Both rate "constants" and equilibrium "constants" are temperature dependent. For example, the equation:

$$\frac{[H^+] \times [F^-]}{[HF]} = 7.0 \times 10^{-4}$$

for the equilibrium between an aqueous solution of hydrofluoric acid and its ions applies at only one temperature, 25°C. At temperatures that differ significantly from 25°C, the equilibrium quotient represented as K_a will have some value other than 7.0×10^{-4}. At 100°C, for example, the equilibrium constant for the dissociation of hydrofluoric acid in water is 2.5×10^{-4}.

The value of the dissociation constant of a weak acid also depends strongly upon the nature of the acid. While it is true that for any weak acid:

$$HA(aq) \rightleftharpoons H^+(aq) + A^-(aq)$$

the expression:

$$\frac{[H^+] \times [A^-]}{[HA]} = K_a$$

is always valid, the numerical value of K_a at 25°C varies widely, being 1.8×10^{-5} for acetic acid, 7.0×10^{-4} for hydrofluoric acid, and 4.0×10^{-10} for HCN, a very weak acid.

In contrast, the ideal gas law, $PV = nRT$ with $R = 0.0821$ liter atm/mole °K, is an extremely general relationship. To a good degree of approximation, it applies to all gases, pure or mixed, over a wide range of experimental conditions.

4. **Constants frequently have units, in which case their magnitude will, of course, depend upon the units in which they are expressed.**

You may recall that in this chapter we have used two different values for the gas constant, R, one in liter atm/mole °K, the other in cal/mole °K. The conversion from one set of units to the other is made readily.

$$R = 0.0821 \frac{\text{liter atm}}{\text{mole °K}} \times \frac{24.2 \text{ cal}}{1 \text{ liter atm}} = 1.99 \frac{\text{cal}}{\text{mole °K}}$$

EXERCISES

1. Evaluate the constants a, b, and c in the following equations.
   a. $y = a/x$; $y = 6$ when $x = 4$.
   b. $y = ax + b$; $y = 6$ when $x = 4$, $y = 2$ when $x = 0$.
   c. $y = a + bx + cx^2$; $y = 1$ when $x = 0$, $y = 2$ when $x = 1$, $y = 1$ when $x = 2$.

2. By taking an average, find the "best value" of the constant a in the equation $y = ax^2$.

y	4.1	16.9	38.0	67.0
x	1.0	2.0	3.0	4.0

PROBLEMS

9.1 The volume of a gas confined at constant temperature and pressure is directly proportional to the number of moles. One mole of a gas occupies 22.4 liters at $0°C$ and 1 atm pressure. Under these conditions:

    a. What is the volume occupied by 1.63 moles?

    b. How many moles are required to occupy 12.0 liters?

9.2 The average velocity, u, of a gas molecule is directly proportional to the square root of the absolute temperature, T, and inversely proportional to the square root of the molar weight, M.

    a. Express these two functional relationships in terms of a single equation (use a to represent the constant).

    b. Write a two-point equation relating u_2 to u_1, T_2, T_1, M_2, and M_1.

9.3 The functional relationship between the volume, V, of a gas and its absolute temperature, T, can be expressed by the equation:

$$V = aT \ (n \text{ and } P \text{ constant})$$

Using the ideal gas law, $PV = nRT$, with $R = 0.0821$ liter atm/mole $°K$, evaluate the constant a in this equation (V in liters, T in $°K$) when:

    a. $n = 1.00$ mole, $P = 1.00$ atm

    b. $n = 5.00$ moles, $P = 0.100$ atm

9.4 Using the relation given in Problem 9.3, show that V is a linear function of the temperature, t, expressed in $°C$.

9.5 The vapor pressure, P, of the solvent in a liquid solution is directly proportional to its mole fraction, X. Complete the following table.

P					160 mm Hg	
X	0.00	0.20	0.40	0.60	0.80	1.00

9.6 The freezing point lowering (fpl) of a solution is directly proportional to the number of grams of solute (g_2), inversely proportional to the number of grams of solvent (g_1), and inversely proportional to the molar weight of the solute (M_2).

    a. Write a single equation expressing these three functional relationships. (Use a to represent the constant.)

    b. The freezing point lowering of a certain solution is $1.00°C$. What will be the freezing point lowering if the number of grams of solute is doubled? if the number of grams of solvent is increased from 10.0 to 15.0?

    c. Evaluate the constant a in the equation, given that a solution containing 1.00 gram of solute of molar weight 60.0 dissolved in 10.0 grams of solvent has a freezing point lowering of $0.464°C$.

9.7 Given that the rate of effusion of a gas is inversely proportional to the square root of its molar weight:

    a. Calculate the ratio of the rates of effusion of CH_4 vs. He.

b. Calculate the molecular weight of a gas which effuses 0.82 times as rapidly as O_2.

9.8 Given the following values for the volume in ml of one gram of mercury as a function of temperature in $°C$, evaluate the constants in the linear equation:

$$V = at + b$$

V (ml)	0.073688	0.074089
t $(°C)$	10.00	40.00

9.9 For the reaction: $2SO_2(g) + O_2(g) \rightarrow 2SO_3(g)$, $\Delta H = -47.0$ kcal and $\Delta G = -33.4$ kcal at $300°K$. Using the equation: $\Delta G = \Delta H - T\Delta S$, calculate
   a. ΔS b. ΔG at $500°K$ c. the temperature at which $\Delta G = 0$.

9.10 Given the following data for the reaction: $2NO(g) + O_2(g) \rightarrow 2NO_2(g)$, show that the ratio: $[NO_2]^2/[NO]^2 \times [O_2]$ is constant and evaluate the constant.

[NO]	[O_2]	[NO_2]
1.00	1.00	3.16
1.00	2.00	4.47
2.00	1.00	6.32
2.00	2.00	8.95

9.11 For the equilibrium between atomic fluorine, F, and molecular fluorine, F_2, the following data is obtained:

[F]	0.316	0.447	0.548
[F_2]	1.00	2.00	3.00

It is believed that the functional relationship is of the form $[F_2] = K[F]^n$, where n may be any positive whole number (i.e., 1, 2, 3 $\cdots$). Use the data given to evaluate n and then K.

9.12 For the reaction: $A + B + C \rightarrow$ products, the rate expression has the form:

$$\text{rate} = k(\text{conc. } A)^m (\text{conc. } B)^n (\text{conc. } C)^p$$

where m, n, and p may each be 0, 1, or 2. Use the following data to evaluate m, n, and p.

rate	conc. A	conc. B	conc. C
0.0050	0.10	0.10	0.10
0.0100	0.20	0.10	0.10
0.0100	0.20	0.20	0.10
0.0400	0.20	0.20	0.20
0.0600	0.30	0.20	0.20
0.1350	0.30	0.20	0.30
0.0600	0.30	0.30	0.20

9.13 For the decomposition of acetaldehyde: rate = $k(\text{conc. } CH_3CHO)^2$

a. What are the units of k if the rate is expressed in mole/liter sec and the concentration in mole/liter?

b. How would the magnitude of k change if the rate were expressed in mole/liter min?

9.14 The vapor pressure of chloroform is given by the equation:

$$\log P = \frac{-1610}{T} + 4.81; \quad \left(P \text{ in atm, } T \text{ in } °K\right)$$

a. What is the vapor pressure of chloroform at 25°C?

b. What is the normal boiling point of chloroform (i.e., the temperature at which $P = 1$ atm)?

c. Using the relation given in Table 9.5, calculate the molar heat of vaporization of chloroform.

9.15 The vapor pressure of water at 50°C is 92.5 mm Hg; at 100°C it is 760 mm Hg. Evaluate the constants in the equation:

$$\log P = \frac{-A}{T} + B; \quad (P \text{ in mm Hg, } T \text{ in } °K)$$

9.16 Consider the following equilibrium expressions. In each case, state how the concentration of the species in bold type is related to that of the other species.

Example: $\dfrac{[Cl_2] \times [PCl_3]}{[PCl_5]} = K$: directly proportional to $[PCl_5]$
inversely proportional to $[PCl_3]$

a. $\dfrac{[\mathbf{H_2}] \times [I_2]}{[HI]^2} = K$

b. $\dfrac{[H_2] \times [I_2]}{[\mathbf{HI}]^2} = K$

c. $[\mathbf{Ag^+}] \times [Cl^-] = K$

d. $\dfrac{[\mathbf{Ag^+}] \times [NH_3]^2}{[Ag(NH_3)_2{}^+]} = K$

e. $\dfrac{[Ag^+] \times [NH_3]^2}{[\mathbf{Ag(NH_3)_2{}^+}]} = K$

f. $\dfrac{[H^+] \times [F^-]}{[\mathbf{HF}]} = K$

GRAPHS

We pointed out in Chapter 9 that a functional relationship between two variables, y and x, can be presented in the form of an algebraic equation, a table giving corresponding values of y and x, or a graph in which y is plotted against x. In that chapter, our discussion was limited to the first two forms; here we will consider graphical methods of describing functional relationships. To begin with, it will be helpful to classify the various types of graphs that you will work with in general chemistry.

10.1 TYPES OF GRAPHS AND HOW TO READ THEM

All graphs found in a general chemistry textbook are two-dimensional in the sense that they are printed on a piece of paper. However, in a more meaningful way, graphs can be classified as one-dimensional, two-dimensional, or three-dimensional, according to the number of variables that are being plotted.

One-Dimensional Graphs. The simplest type of graph is one in which only a single variable is plotted. Such graphs are referred to as one-dimensional, since only distances in one dimension, either horizontal or vertical, have meaning. Two examples of the most common type of one-dimensional graph in general chemistry are shown in Figure 10.1. In both cases, a single variable (oxidation number in 10.1a, heat of formation in 10.1b) is plotted along a vertical line. Values assigned to various compounds of nitrogen are indicated by arrows to the right of the line.

Graphs of this type are relatively easy to interpret. We see, for example, from Figure 10.1a that the oxidation number of nitrogen in NH_3 is -3. From Figure 10.1b, it appears that the heat of formation of NH_3 is about -11 kcal/mole, since the arrow for that compound is located about one-fifth of the way from -10 kcal/mole to -15 kcal/mole.

Sometimes, the second dimension in a one-dimensional graph carries at least a

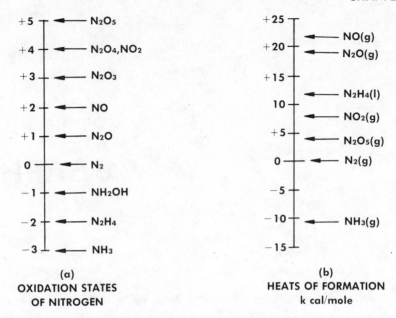

(a)
OXIDATION STATES
OF NITROGEN

(b)
HEATS OF FORMATION
k cal/mole

FIGURE 10.1 One-dimensional graphs.

vague qualitative meaning. In Figure 10.2, we show a typical activation energy diagram designed to illustrate energy changes in the reaction:

$$A \rightarrow B$$

The horizontal line at the left represents the energy of the reactant molecule, A; that at the far right gives the energy of the product molecule, B. The high point in the middle of the graph represents the energy of the "activated complex," A^*, an unstable intermediate formed in the course of the reaction. In a general sort of way, distance in the horizontal direction represents time. We start with pure A, pass through the intermediate A^*, and arrive finally at B. However, the horizontal distances have no quantitative meaning; we have no idea how long it takes to form A^* from A or B from A.*

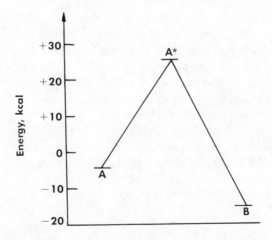

FIGURE 10.2 Activation energy diagram.

Example 10.1 The activation energy for the reaction shown in Figure 10.2 is defined as $E_{A*} - E_A$. Estimate the activation energy from the graph.

Solution It appears that E_{A*} is about +25 kcal (approximately half way between +20 and +30 kcal). Again, E_A seems to be about −4 kcal (a little less than half way from 0 to −10 kcal). Thus:

activation energy = 25 kcal − (−4 kcal) = +29 kcal

Considering the uncertainty in reading the graph at both locations to be of the order of ±1 kcal, we should not be surprised if our answer is off by as much as 2 kcal.

Two-Dimensional Graphs. By far the most common type of graph in chemistry (and the only type discussed in the remaining sections of this chapter) is a two-dimensional graph, used to show the functional relationship between two variables, y and x. Although you have probably seen and interpreted many such graphs, it may be helpful to review the general principles involved by referring to the particularly simple y-x graph shown in Figure 10.3. You will note that the background consists of a grid of parallel and perpendicular lines. Superimposed on this grid are a series of points lettered A through E. Each point represents a pair of correspond-

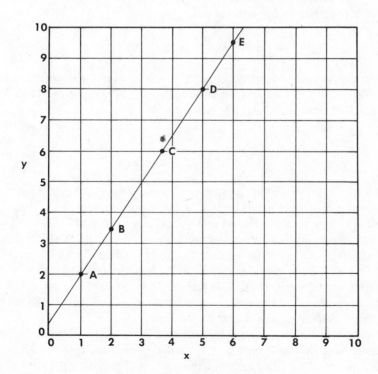

FIGURE 10.3 A simple straight-line graph.

ing values of y and x. The graph itself is the smooth curve (in this case, a straight line) drawn through the several points.

In Figure 10.3, as in all two-dimensional graphs, values of the independent variable, x, are shown below the horizontal axis, increasing in magnitude from left to right (0 at the far left, 10 at the far right). Any point with an x value of 2 must be 2 units out from the vertical axis and hence must fall somewhere on the vertical line numbered "2" at the bottom. Similarly, any point for which $x = 6$ must lie on the vertical line drawn six units to the right of the vertical axis, i.e., the line labeled "6" at the bottom.

Values of the dependent variable y are indicated to the left of the vertical axis, increasing from bottom (0) to top (10). The horizontal lines on the grid represent constant values of y. Any point lying on the horizontal line labeled "3" at the far left, located three units up from the horizontal axis, must have a y value of 3. Similarly, any point for which $y = 10$ must fall somewhere along the horizontal line drawn 10 units up from the horizontal base line and labeled "10" at the far left.

To assign x and y values to the points shown in Figure 10.3, we need only consider their positions on the grid of vertical and horizontal lines. In a particularly simple case, we see that point A is at the intersection of the vertical line numbered "1" and the horizontal line numbered "2." It must then have an x value of 1 and a y value of 2.

$$\text{Point } A: \quad x = 1, y = 2$$

To assign values to point B, we first note that it falls exactly on the line corresponding to $x = 2$. It appears to be about half way from the line $y = 3$ to the line $y = 4$. Thus, a reasonable assignment would be:

$$\text{Point } B: \quad x = 2, y = 3.5$$

For point C, the y value is obviously 6; it is exactly on the horizontal line six units up from the axis. The x value is not so easily determined; it appears to be somewhat more than half way from the vertical line $x = 3$ to the vertical line $x = 4$; a reasonable estimate might be 3.7.

$$\text{Point } C: \quad x = 3.7, y = 6$$

In a similar manner, we could assign x and y values to points D and E (see Exercise 2) or indeed to any other points located on the straight-line graph of Figure 10.3.

Most of the graphs that you will find in a chemistry text differ from Figure 10.3 in one important respect: the background grid is omitted, although the graph is interpreted as though the grid were present. Constant values of x and y are indicated, not by vertical and horizontal lines, but by hash marks drawn along the horizontal and vertical axes (Figure 10.4). This marking system has the advantage of making the graph itself stand out more clearly; unfortunately it also makes it harder to pick out corresponding values of x and y.

To assign x and y values to a point on a graph such as that shown in Figure 10.4, it is helpful to draw dotted lines from the point, perpendicular to the two axes. By observing where these lines intersect the axes, it is possible to locate the point

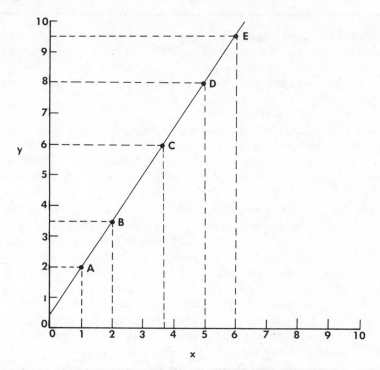

FIGURE 10.4 Graph of Figure 10.3 without grid.

within fairly narrow limits. Consider, for example, point A in Figure 10.4. The dotted vertical line appears to intersect the x axis at about $x = 1$; the dotted horizontal line intersects the y axis at the hash mark corresponding to $y = 2$. (This position is hardly surprising, since the graph in Figure 10.4 was drawn to be identical with that in Figure 10.3, where we found the x and y values of point A to be 1 and 2, respectively). Proceeding similarly with points B and C, we deduce that the x and y values are approximately:

	x	y
Point B	2	3.5
Point C	3.7	6

Note, for example, that the horizontal dotted line from point B intersects the y axis about midway between the hash marks labeled "3" and "4." Again, the vertical dotted line from point C cuts the x axis somewhere between 3 and 4, perhaps about seven-tenths of the way from 3 to 4.

Frequently, with graphs of this type, we find that only certain of the hash marks along the axes are numbered. In Figure 10.4, for example, we might find only the divisions 0, 5, and 10 marked. When we find this type of marking, we always assume that the unnumbered divisions are evenly spaced. Thus, finding four unnumbered marks between $x = 0$ and $x = 5$, we take them to represent $x = 1, x = 2,$ $x = 3,$ and $x = 4.$

Example 10.2 Using Figure 10.5, which shows the vapor pressure of water as a function of temperature, estimate:
  a. The vapor pressure of water at 25°C.
  b. The temperature at which the vapor pressure of water is 60 mm Hg.

Solution Notice that along both axes, the numbered marks are at intervals of 10 units (i.e., 10°C, 10 mm Hg). The small hash marks midway between the larger ones represent intervals of 5 units. For example, the mark midway between 20 and 30 on the horizontal axis represents 25°C.
  a. Here, we first draw a dotted vertical line up to the curve, starting at 25°C on the horizontal axis. At the point of intersection with the curve, we draw another line horizontally across to the y axis. The intersection of this line with the y axis appears to lie slightly below the small hash mark at 25 mm Hg. A good estimate would be 24 mm Hg.
  b. Here, we reverse the procedure in (a), first extending the mark at 60 mm Hg across to the curve, then dropping a perpendicular down to the x axis. The intersection of this line with the axis lies slightly beyond 40°C, perhaps at 41°C.

The major source of error in reading graphs of this type is failure to draw the dotted lines at angles of 90° to the axes. If you are really fussy about it, you can use a transparent plastic triangle or perhaps even a T-square to draw these lines. However, it is seldom worth the effort to do this. Unruled graphs are designed to

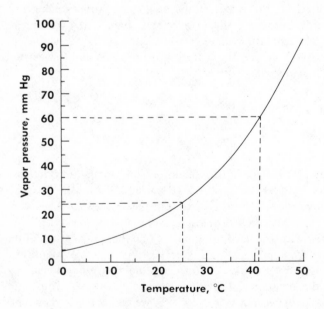

FIGURE 10.5 Vapor pressure of water, 0 to 50° C.

show the general form of a functional relationship as clearly as possible; even when drawn very carefully, they do not lend themselves to highly accurate readings.

Three-Dimensional Graphs. If we wish to show the functional relationship between three variables, x, y, and z, we could construct a solid, three-dimensional model. More commonly, graphs of this type are drawn in perspective on paper. The three axes, x, y, and z, are considered to be at $90°$ angles to each other. In Figure 10.6, we show the z axis running north and south, the x axis running east and west, and the y axis perpendicular to the plane of the paper. A point drawn on such a graph represents a set of three corresponding values of x, y, and z. A group of such points traces out a graph that is supposed to represent a three-dimensional figure.

Graphs involving three variables occur relatively seldom in general chemistry. One area in which they are helpful is in describing the "shapes" of atomic orbitals. Figure 10.6 represents the three p orbitals (p_x, p_z, and p_y). Specifically, the fuzzy, twin lobes enclose regions of space in which there is a high probability of finding a p electron. Quite obviously, this graph gives us very little quantitative information. Indeed, we have not even attempted to indicate absolute distances along the three axes. Qualitatively, however, Figure 10.6 tells us at a glance that the three p orbitals have identical shapes, are oriented at 90 degree angles to each other, and consist of two identical lobes, one on each side of the axis. Information of this sort can be quite useful in making predictions about the electronic structures and geometries of molecules.

EXERCISES

1. Referring to Figure 10.1, what is the oxidation number of nitrogen in NO? the heat of formation of NO?
2. Referring first to Figure 10.4 and then to Figure 10.3, give the values of y and x corresponding to points D and E.
3. Referring to Figure 10.5:
   a. What is the vapor pressure of water at $10°C$?
   b. At what temperature is the vapor pressure 30 mm Hg?

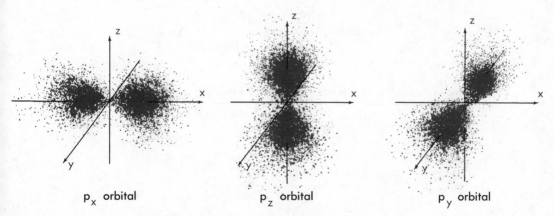

p_x orbital p_z orbital p_y orbital

FIGURE 10.6 Electron clouds associated with p orbitals.

10.2 GRAPHING DATA

In Section 10.1, we saw how to read a two-dimensional graph. Basically, what we did was to generate a set of x and y values corresponding to points on the graph. Here, we consider the reverse problem, the construction of a graph from a table of experimental data expressing y as a function of x. Although graph paper comes in a variety of forms ruled in different ways, we will assume throughout this section that we are working with one particular type (Figure 10.7 p. 154) which has:

    a. 10 major divisions (heavy lines) and 50 small divisions along the vertical axis.

    b. 7 major divisions (heavy lines) and 35 small divisions along the horizontal axis.

There are two basic principles that guide us in drawing any graph. In order of priority, they are:

    1. The graph should be clear, easy to read, and easy to construct. In particular, it should be possible to determine y and x values of a point with a minimum of effort and see immediately what they represent (e.g., pressure in atmospheres, temperature in $^\circ K$).

    2. The graph should cover most of the graph paper rather than being squeezed into a tiny area. If the points on a graph are very close together, it loses much of its usefulness.

To see how these principles work out in practice, let us construct a graph from the data in Table 10.1, which gives the solubility of tartaric acid as a function of temperature. We will use the horizontal axis to show temperature (the independent variable) and the vertical axis to represent solubility (the dependent variable). In drawing this or any other graph, it is helpful to follow the logical sequence of steps outlined below.

Steps in Drawing a Graph

    1. **Decide how many units each division along the two axes will represent.** Let's consider the horizontal axis first. Remember that we have 7 large divisions along this axis. The temperatures in Table 10.1 run from $20^\circ C$ to $80^\circ C$, a range of 60°. If we insist upon making the graph cover the entire paper, the horizontal axis would start at 20° at the far left and go to 80° at the far right. This would require that each major division correspond to:

$$\frac{60^\circ}{7} = 8.57 \cdots$$

Table 10.1 Solubility of Tartaric Acid (g/100 g water) at Various Temperatures ($^\circ$C)

Point	Solubility	Temperature
1	18	20
2	25	30
3	37	40
4	50	50
5	65	60
6	81	70
7	98	80

Each minor division (5 per major division) would represent:

$$\frac{8.57°}{5} = 1.71 \cdots$$

These numbers would be awkward to work with, to say the least. It would be more sensible to make each major division along the horizontal axis 10°, in which case each minor division would represent $10°/5 = 2°$.

On the vertical axis, we need to cover a range of 80 solubility units (98 g/100 g of water at the highest temperature, 18 g/100 g of water at the lowest temperature) within 10 major divisions. We must assign to each major division at least:

$$\frac{80}{10} = 8$$

solubility units. The graph will be somewhat easier to read if we make each major division along the vertical axis represent 10 units rather than 8; each small division will then represent 2 units rather than 1.6.

Summarizing the reasoning process that we have just gone through, we first divide the range to be covered (60°C, 80 solubility units) by the number of large divisions available along each axis (7, 10). The quotients thus obtained give us the minimum number of units that can be assigned to each major division. If we were to choose these values, we would make the fullest use of the paper, spreading the graph over the total area. Ordinarily, we find that increasing these values somewhat to some convenient integral values leads to a graph that is easier both to construct and to read.

2. **Decide upon the values to be assigned to the point at the lower left corner of the graph paper, commonly called the origin.** An obvious possibility here would be to make point 1 ($y = 18, x = 20$) fall at the origin. As a matter of fact, 20 is a quite reasonable choice for the x value at the origin. The major divisions along the x axis will then be 20°, 30°, 40°, 50°, 60°, 70°, 80°, and 90°. Each of the temperatures listed in Table 10.1 will then fall at a major division, a very happy situation indeed.

On the other hand, 18 would not be a particularly convenient choice for the y value at the origin. The major divisions along the y axis would then read 18, 28, 38, $\cdots$ 118. It would be a little easier to locate y values if we made the major divisions correspond to integral multiples of 10. There are a couple of ways to do this, but perhaps the simplest is to choose a value of 0 for the y origin. The major divisions along the y axis will now read 0, 10, 20, $\cdots$ 100. This system of division neatly covers the range of solubilities, 18 through 98 g/100 g water.

In general, to select appropriate values for the origin, we first look at the smallest y and x values in the table (e.g., 18, 20). Obviously, the origin must be set low enough to include these values on the graph paper; in the example we are discussing, it could not, for instance, be set at $y = 20, x = 30$. Frequently, we will find the graph easier to construct and interpret if we set the origin somewhat lower (e.g., 0, 20). In doing this, we must be careful not to set the origin so low that high values of x and y fall off the paper. If, with the data in Table 10.1, we were to choose 0, 0 for the origin, keeping each major division at 10 units, the x axis would run from 0 to 70, and the highest temperature, 80°, would be pushed off the right edge of the paper.

3. **Number each major division along both axes.** If steps 1 and 2 have been carried out thoughtfully, this one is routine. If not, you may get an unpleasant surprise (see the preceding discussion).

4. **Label both axes, indicating the quantity being plotted.** This procedure has to be carried out eventually, and now is as good a time as any. Be sure to include not only the identity of the variable (solubility, temperature) but also the units in which it is expressed (g/100 g water, °C).

In Figure 10.7, we show the results of Steps 1 to 4 for the case we are studying. Notice that the body of the graph paper is still untouched. Now, at long last, we are ready to construct the graph itself.

5. **Plot the points.** Here, we will consider only points 1 and 2. To find where point 1 belongs, we first locate "18" on the vertical axis. Clearly, it falls between

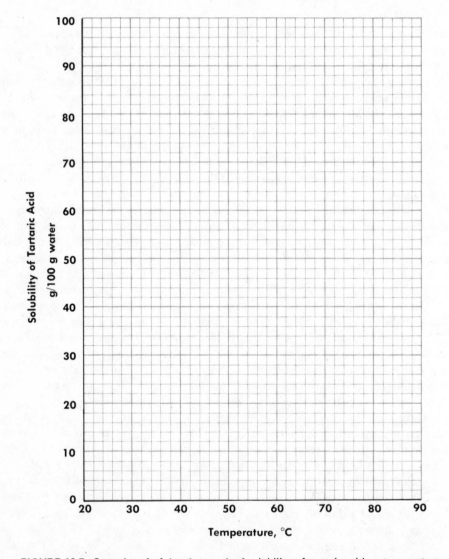

FIGURE 10.7 Steps 1 to 4 of drawing graph of solubility of tartaric acid vs. temperature.

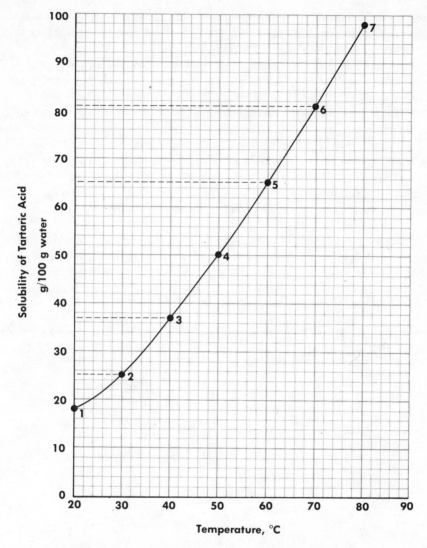

FIGURE 10.8 Graph of solubility of tartaric acid vs. temperature.

the major divisions numbered "10" and "20." To pin it down more closely, we note that each small division represents 2 solubility units. We conclude that a y value of 18 must fall one small division below 20. Keeping this in mind, we try to locate an x value of "20." As it happens, we don't have far to look. Since we chose to have temperatures start at 20° at the far left, any point for which $x = 20$ must fall on the heavy line at the left edge of the paper. In other words, the point $y = 18$, $x = 20$ falls on the y axis, 18 units up from the origin. This point is indicated on Figure 10.8 as a heavy dot marked "1."

To locate point 2, we first have to find where "25" falls on the y axis. Since 25 is midway between 20 and 30, it must be 2 1/2 small divisions above 20. To put it another way, 25 lies half way between the small divisions corresponding to 24 and 26 solubility units. You may find it helpful to enter a light hash mark at this point on the y axis. Now, we move out from this mark, parallel to the x axis, until we come to the line corresponding to $x = 30$, one major division out from the y axis. At the intersection $y = 25, x = 30$, we indicate point 2 with a dot.

Points 3 through 7 are located similarly. Point 4, for example, is located at the intersection of the lines $x = 50, y = 50$. You may wish to verify that we have indeed positioned all of these points properly.

6. **Draw a *smooth* curve through the points.** If we were lucky enough to have all the points fall on a straight line, this step would be very simple indeed. Here, the plot is obviously not linear, at least in the lower portions. A French curve (Figure 10.9) or, even better, a flexible spline is very helpful in drawing curves such as this. Incidentally, in using a French curve, you should avoid the common mistake of trying to draw too large a portion of the curve at once. You will ordinarily have to shift the device several times to avoid getting sharp breaks in the curve.

Comments on Graphing. Sometimes, in attempting to draw graphs to represent experimental data, you will run into problems that were not covered explicitly in the example just discussed. In particular, you may find that the first and most critical step, deciding how many units should be assigned to each major division along an axis, is a bit tricky (Example 10.3).

Example 10.3 Suppose you were asked to plot the solubility (y) of potassium nitrate in water vs. temperature (x), using the following data:

y (g/100 g water)	13	21	32	46	64	86	110	138
x (°C)	0	10	20	30	40	50	60	70

How many units should be assigned to each major division along the two axes?

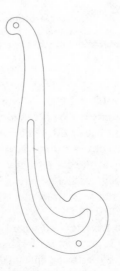

FIGURE 10.9 French curve.

Solution Here, it is very easy to make a decision for the x axis. The range of x values is 70 (i.e., $70° - 0° = 70°$). Recalling that there are 7 major divisions along the x axis, the minimum number of units per division is 10.

$$x \text{ axis:} \quad \frac{70}{7} = 10$$

Clearly, it will be convenient to make each major division count as $10°$, thereby making optimum use of the horizontal dimension of the graph paper.

Where the y axis is concerned, the proper choice is not nearly so obvious. The range of y values is 125 (i.e., 138 g/100 g water – 13 g/ 100 g water = 125 g/100 g water). Dividing by the 10 major divisions along the y axis, we obtain:

$$y \text{ axis:} \quad \frac{125}{10} = 12.5$$

as the minimum number of units to be assigned to each major division. Clearly, for convenience in graphing, this number should be increased. The question is: how far should we go? If we were to increase it to 20, we would be using only a little more than half of the vertical dimension, which seems rather wasteful. At the opposite extreme, we might choose 13 units per major division. However, 13 would not be much better than 12.5; each small division would now be $13/5 = 2.6$ units, an awkward choice.

A reasonable compromise would be to make each major division along the y axis represent 15 solubility units. With this choice, we can start at 0 at the origin and proceed in units of 15 up to 150 at the top of the paper. Most important, each small division represents an integral number of units: $15/5 = 3$. This choice makes it quite easy to locate points on the solubility graph (Figure 10.10, p. 158).

A minor but annoying problem arises when the numbers we have to work with are either very large (e.g., 1,000,000) or very small (e.g., 0.000001). How should we label the major divisions of the axes in such cases? Writing out the numbers would consume so much space that the labels would run together and be hard to read. Two approaches may be used here.

a. The numbers are expressed in exponential notation (e.g., 1×10^6, 1×10^{-6}) and written directly along the axes. This notation is clear enough but unfortunately doesn't save much space. Notice, for example, that there are still five characters in 1×10^6 and six in 1×10^{-6}.

b. A number such as 1,000,000 is indicated on the axis simply as "1." Down below, the axis itself is labeled as "$y \times 10^{-6}$." This solves the space problem, but can be extremely confusing, at least the first time you see it. The understanding is that:

$$y \times 10^{-6} = 1; \quad \text{hence } y = 1/10^{-6} = 1,000,000$$

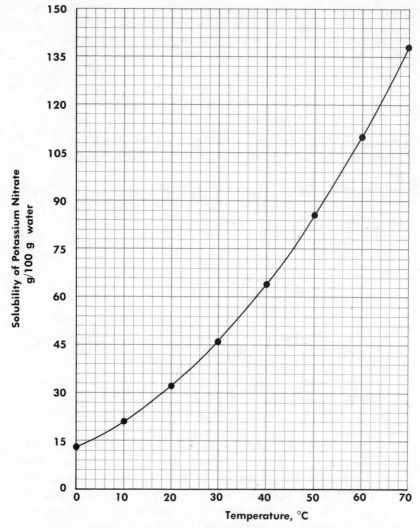

FIGURE 10.10 Graph of solubility of potassium nitrate versus temperature.

In this system, an x value of 0.000001 would be indicated on the axis as "1" with the notation underneath that the quantity being plotted is $x \times 10^6$.

$$x \times 10^6 = 1; \quad x = 1/10^6 = 0.000001$$

In graphing experimental data, it is desirable to indicate in some way the precision of the measurements from which the data were obtained. One way to do this is to adjust the size of the dots on the graph. Small points suggest data of high precision; large points indicate relatively crude data. A more meaningful approach is to indicate the number of significant figures in the measurements by the way in which divisions along the axis are numbered. For example, referring back to Figure 10.8, if the solubility measurements had been made to ±0.1 gram, we would write the numbers 20.0, 40.0 ··· along the y axis instead of 20, 40 ··· .

With data of low precision, we frequently find that one or more points appear to be significantly off the curve. This effect is most apparent when we are dealing with a straight-line graph (see Section 10.4); it is less obvious but equally serious if the plot is curved. If there is good reason to believe that a particular measurement is in error, it is legitimate to ignore that point in constructing the graph. Otherwise, you should use all the points and draw a smooth curve that comes as close as possible to each point. Under no circumstances should you make a "zigzag" plot in an attempt to force the curve to pass through every point.

EXERCISES

1. Consider the following sets of data. In each case, decide how many units should be represented by each major division on your graph paper, choose appropriate numbers for the origin, and plot the data.

   a. y 5 7 9 11 13 15
      x 1 2 3   4   5   6

   b. y 3.0 1.5 1.0 0.75 0.60 0.50
      x 1.0 2.0 3.0 4.0   5.0   6.0

   c. y 5.0   7.0   9.0   11.0   13.0
      x 0.20 0.40 0.60 0.80   1.00

   d. y 0   7.1   10.0   12.3   14.4
      x 0 50   100   150   200

10.3 GRAPHING EQUATIONS

To construct the graph corresponding to an algebraic equation, we follow essentially the same procedure outlined in the previous section, with one additional step. We must first use the equation to generate a series of data points.

Example 10.4 Given the equation:

$$^\circ F = 1.8\,^\circ C + 32^\circ$$

construct a plot of $^\circ F$ (y axis) vs. $^\circ C$ (x axis), covering the range 0 to 100°C.

Solution Along the x axis, it will be convenient to start with 0°C at the origin and let each major division count 20°. With this method of numbering, it seems reasonable to calculate values of $^\circ F$ corresponding to 0°, 20°, 40°, 60°, 80°, and 100°C. Using the equation given, we arrive at the following set of data points.

$^\circ F$ 32 68 104 140 176 212

$^\circ C$ 0 20   40   60   80 100

We must now consider what numbers to assign to major divisions along the y axis. Note that the range is $212° - 32° = 180°$. We see that each of the 10 major divisions must represent at least $180°/10 = 18°$. It would probably simplify matters to increase this number to $20°$. Our first impulse might be to let the y origin be $0°$, like the x origin. However, this choice clearly will not work. The top line on the paper would then represent $200°F$; the highest temperature, $212°F$, would fall off the grid. A simple way to avoid this problem is to start at $20°$ on the y axis. The heavy lines corresponding to major divisions along that axis will then run from $20°$ to $220°$ at $20°$ intervals.

Having gone this far, you only need to plot the points in the usual way and draw a smooth curve through them. The results are shown in Figure 10.11; note that the "smooth curve" is actually a straight line.

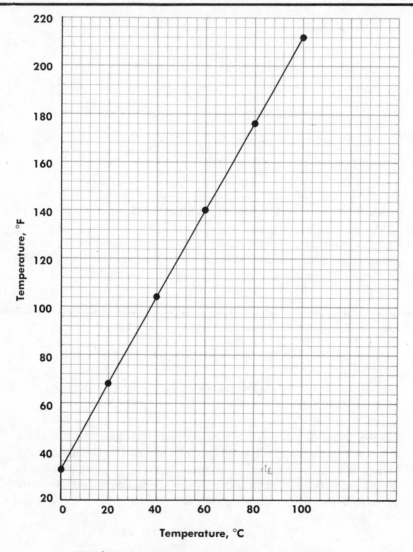

FIGURE 10.11. Graph of °F vs. °C from 0 to 100°C.

As Example 10.4 implies, it is ordinarily sufficient, in graphing an equation, to generate one data point for each major division along the x axis. As a matter of fact, we can get by with fewer points than that if we are graphing a linear function, i.e., if the equation is of the form:

$$y = ax + b$$

Since the position of a straight line is determined by two points, we need only generate points at the beginning and end of the range. In Example 10.4, the two points:

°F	32	212
°C	0	100

would have been sufficient.

In graphing certain equations, we may find it helpful to generate additional data points in order to be more certain about the location of the curve in a particular region (Example 10.5).

Example 10.5 For one mole of an ideal gas at 92°C, the relation between volume and pressure is given by:

$$V = 30.0/P \quad (V \text{ in liters, } P \text{ in atm})$$

Plot V (y axis) vs. P (x axis) from $P = 3$ atm to $P = 10$ atm.

Solution With 7 major divisions along the x axis, it is clearly convenient to let each division represent 1 atm.. We next calculate V at pressure intervals of 1 atm from 3 to 10 atm.

V(liters)	10.0	7.50	6.00	5.00	4.29	3.75	3.33	3.00
P(atm)	3.0	4.0	5.0	6.0	7.0	8.0	9.0	10.0

Looking at the values of V, we see that they cover a range of 10.0 − 3.0 = 7.0 liters. It seems reasonable in this case to make the y axis symmetrical with the x axis, i.e., we start with $V = 3$ liters at the origin and let each major division represent 1 liter.

In Figure 10.12, the points generated above are plotted as solid dots. As you can see, V is decreasing rapidly near the beginning of the curve, particularly in the region between 3 and 4 atmospheres. To construct an accurate curve, it might help to have another point in this region, perhaps at 3.5 atm. Just to be safe, we might obtain another point at 4.5 atm.

$$P = 3.5 \text{ atm}; \quad V = 30.0/3.5 = 8.57 \text{ liters}$$

$$P = 4.5 \text{ atm}; \quad V = 30.0/4.5 = 6.67 \text{ liters}$$

These two points are shown as open circles in Figure 10.12. As you can see, they appear to fill in gaps in the plot, thereby locating the curve more reliably in this region.

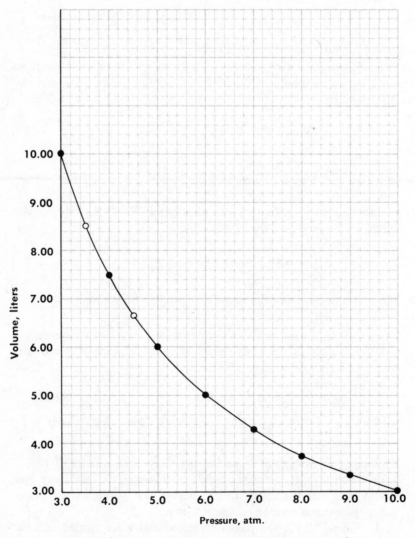

FIGURE 10.12 P vs. V graph for one mole of ideal gas at 92° C.

Frequently, when asked to graph an equation, we are not expected to make an accurate plot on a piece of graph paper. A crude sketch indicating the general nature of the curve may be sufficient. In Figure 10.13, we show such sketches for the various functional relationships discussed in Chapter 9. Two notes of caution are in order:

1. In these sketches, only positive values of y and x are shown; the origin is taken to be $(0,0)$. Negative values are possible, though less common in chemistry.

2. In constructing these graphs, the constants a and b were taken to be positive numbers, which is perhaps most frequently the case in equations used in general chemistry. The graphs would look quite different if a or b were negative (see Section 10.4).

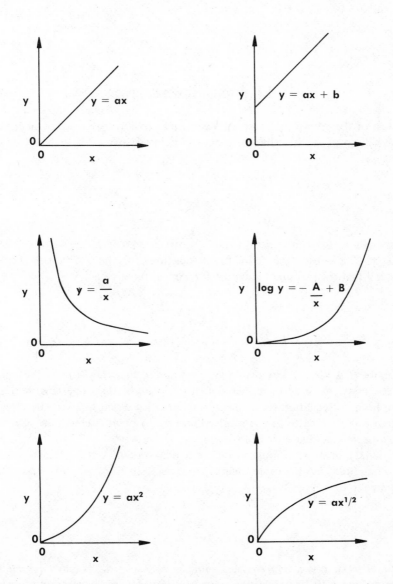

FIGURE 10.13 Typical graphs of functional relationships (all constants taken to be positive).

EXERCISES

1. Draw graphs for the following equations in the regions indicated.

   a. $y = 3x + 2$; $x = 0$ to $x = 6$

   b. $y = 2x^2$; $x = 0$ to $x = 7$

   c. $\log y = \dfrac{-3}{x} + 2$; $x = 0$ to $x = 2$

   d. $y = 5/x$; $x = 1$ to $x = 6$

2. Consider the equation: $y = x + b$. Draw, on the same paper, graphs covering the region $x = 0$ to $x = 7$ for:

   a. $b = 1$ b. $b = 0$ c. $b = -1$

10.4 STRAIGHT-LINE GRAPHS

Most of the graphs that you will be asked to construct in the general chemistry course will be of the straight line type. As pointed out earlier, an equation of the form:

$$y = ax + b$$

corresponds to a linear function and hence gives a straight line when y is plotted against x. The constants a and b fix the position of the line. You may recall from Chapter 9 that the two-point equation for a linear function is:

$$\frac{y_2 - y_1}{x_2 - x_1} = \frac{\Delta y}{\Delta x} = a$$

The quotient $\Delta y/\Delta x$ is, by definition, the slope of the straight line. We conclude that the constant, a, gives us the numerical value of the **slope** (Figure 10.14a). Again, from the equation for a linear function, we see that when $x = 0, y = b$. The constant b is referred to as the **y intercept**, i.e., the point at which the straight line crosses the x axis (Figure 10.14b).

A variety of more complex functional relationships can be transformed into linear functions by a simple change in variables. Consider, for example, the function:

$$y = a/x$$

A plot of y vs. x is a hyperbola; a typical example is the pressure-volume graph shown in Figure 10.12. However, if we plot y vs. $1/x$, we get a straight line with a

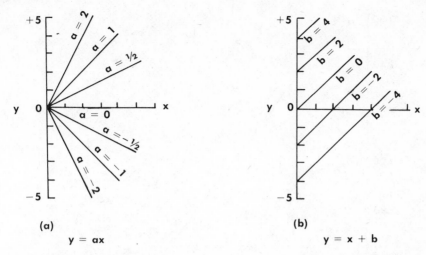

FIGURE 10.14 Effect of constants a (slope) and b (y intercept) on position of linear graph.

slope of *a* (Figure 10.15). In other words, by changing the independent variable from *x* to 1/*x*, we transform a hyperbola into a linear function.

Another functional relationship which is readily converted to a linear function is:

$$\log y = \frac{-A}{x} + B$$

If we choose log *y* and 1/*x* to be our variables rather than *y* and *x*, we obtain a straight line (Figure 10.16, p. 166).

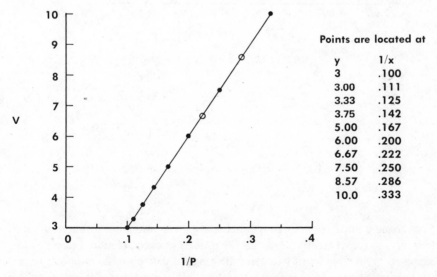

Points are located at	
y	1/x
3	.100
3.00	.111
3.33	.125
3.75	.142
5.00	.167
6.00	.200
6.67	.222
7.50	.250
8.57	.286
10.0	.333

FIGURE 10.15 Conversion of hyperbola into a straight line. The points plotted are those shown in Figure 10.12.

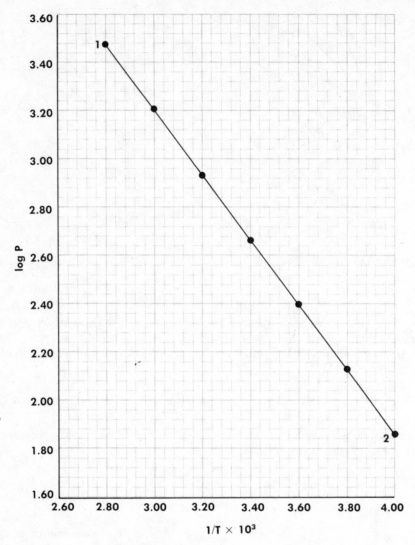

FIGURE 10.16   Log P vs. 1/T for ether.

Converting a more complex functional relationship into a linear function has an obvious advantage. It greatly simplifies the process of **extrapolation**, i.e., extending the graph to predict values of y beyond the range of data plotted. Suppose, for example, that we wanted to use Figure 10.12 to find the volume corresponding to a pressure of 2.0 atm. This would be rather difficult to do, since it would require extending the curve in a region where the slope is changing rapidly. On the other hand, it would be relatively easy to extend the straight line in Figure 10.15 to find that:

$$V = 15.0 \text{ when } P = 2.0 \quad (1/P = 0.50)$$

Interpolation, reading values of y between data points on the curve, is also made easier by a change in variables that converts a curve into a straight line. There is always some uncertainty about how a curve should be drawn through successive points, particularly if they are far apart. If the curve we draw does indeed wander off the true course, any values read from it will be in error. We have more confidence in our ability to locate the position of a straight line and hence in the values we read from it.

Transforming complex functions to linear functions has still another advantage. It often happens that the slope (and sometimes the intercept) of the resultant straight line has direct physical significance. Consider, for example, the relationship between the vapor pressure of a liquid and the absolute temperature (Table 9.5, Chapter 9).

$$\log P = -\frac{\Delta H_{vap}}{2.30\,RT} + B$$

If $\log P$ is plotted against $1/T$, as in Figure 10.16:

$$\text{slope} = \frac{-\Delta H_{vap}}{(2.30)(1.99)}$$

This equation allows us to calculate the heat of vaporization of a liquid from a plot of $\log P$ vs. $1/T$.

Example 10.6 Estimate the heat of vaporization of ether from Figure 10.16.

Solution Perhaps the most accurate way to evaluate the slope of a straight line is to work with the points at the ends of the line. For point 2:

$$x_2 : 1/T_2 \times 10^3 = 4.00; \quad 1/T_2 = 4.00 \times 10^{-3}$$

$$y_2 : \log P_2 = 1.86$$

For point (1):

$$x_1 : 1/T_2 \times 10^3 = 2.80; \quad 1/T_2 = 2.80 \times 10^{-3}$$

$$y_1 : \log P_2 = 3.48$$

$$\text{Slope} = \frac{y_2 - y_1}{x_2 - x_1} = \frac{1.86 - 3.48}{(4.00 - 2.80) \times 10^{-3}} = \frac{-1.62 \times 10^3}{1.20} = -1.35 \times 10^3$$

(Note that the slope is negative; y decreases when x increases.)
To evaluate ΔH_{vap}:

$$\Delta H_{vap} = -(2.30)(1.99)(-1.35 \times 10^3)\, cal = 6180\, cal$$

Ordinarily, when we plot the data points corresponding to a linear function, we find that it is impossible to draw a straight line that passes exactly through all of the points. At best, some of the points will be above the line and some will be below it. What we attempt to do is to draw the line in such a way that we have about as many points above the line as below. More exactly, we estimate visually the position of the line such that the sum of the distances of points above it will be equal to the sum of the distances of points below the line. A transparent plastic straightedge, triangle, or ruler should be used, so that we can see all the points while we are trying to decide where to draw the line.

Example 10.7 Given the following data for the free energy change of a reaction, ΔG, as a function of temperature:

$$\Delta G\ (kcal)\quad 2.8\quad 4.8\quad 6.0\quad 7.7\quad 9.4$$
$$T\ (^\circ K)\qquad\quad 100\ \ 200\ \ 300\ \ 400\ \ 500$$

Use a graphical method to determine the constants ΔH (enthalpy change) and ΔS (entropy change) in the equation:

$$\Delta G = \Delta H - T\Delta S$$

Solution In Figure 10.17, we have plotted ΔG vs. T; the five data points are shown as small circles. We have attempted to draw the "best" straight line through these points. Notice that the points at $200^\circ K$ and $500^\circ K$ fall slightly above the line, while those at $100^\circ K$ and $300^\circ K$ fall slightly below it. It appears that the sums of the distances above and below the line are about equal.

Comparing the equation given to the general linear function:

$$y = ax + b$$

we see that the intercept is ΔH and the slope is $-\Delta S$. The y intercept we can read from the graph as being about +1.5 kcal. Thus:

$$\Delta H = +1.5\ kcal$$

To obtain the slope, we evaluate ΔG and T at the ends of the line.

upper end: $\Delta G = +10.0\ kcal,\ T = 550^\circ K$

lower end: $\Delta G =\ \ +1.5\ kcal,\ T = 0^\circ K$

$$slope = \frac{(10.0 - 1.5)\ kcal}{(550 - 0)^\circ K} = \frac{8.5\ kcal}{550^\circ K} = 0.015\ kcal/^\circ K$$

$$\Delta S = -0.015\ kcal/^\circ K$$

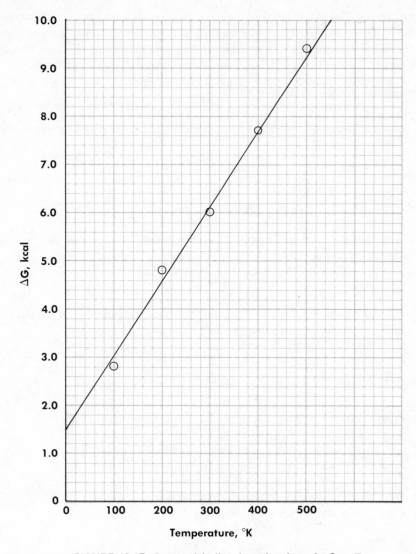

FIGURE 10.17 Best straight line through points of ΔG vs. T.

Attempting to draw the best straight line through a series of data points can be a somewhat uncertain process if there is considerable scatter in the points. An alternative way of determining the constants a and b in the equation:

$$y = ax + b$$

where many points are involved is to use a statistical approach known as the method of *least squares*. As it happens, this method is particularly suitable for computer analysis. We shall not attempt to describe least squares calculations here; an elementary treatment of the method can be found in any of the following references.

H. W. Salzberg, J. I. Morrow, and S. R. Cohen: *Laboratory Course in Physical Chemistry*, Academic Press, New York, 1966, pp. 23–27.

W. J. Youden: *Statistical Methods for Chemists*, Wiley and Sons, Inc., New York, 1951, Chapter 5.

F. Daniels: *Mathematical Preparation for Physical Chemistry*, McGraw-Hill Paperbacks, New York, p. 237.

EXERCISES

1. What quantities should be plotted to give straight line graphs with the following functions?

   a. $\log y = ax + b$ c. $y = ax + bx^2$
   b. $y = ax^2$

2. Using a graphical approach, determine the slope and intercept of the straight lines obtained from the following data.

   a. y -4.0 -2.0 0.0 2.0 4.0
   x 1.0   2.0 3.0 4.0 5.0

   b. y 3.1   5.8 8.9 12.2 14.8
   x 1.0   2.0 3.0   4.0   5.0

   c. y 1.5   2.1 2.6 3.0 3.6
   x 1.0   2.0 3.0 4.0 5.0

   d. y 0.8 -1.2 -3.4 -5.3 -7.5
   x 1.0   2.0   3.0   4.0   5.0

3. Assuming that the following data can be fitted to the equation: $\log y = \dfrac{-A}{x} + B$, determine A and B by a graphical method.

   y 0.011 0.17 0.50 0.76
   x 1.0   2.0   3.0   4.0

PROBLEMS

10.1 Construct a one-dimensional graph to show the following standard reduction potentials.

	SRP	REACTION
$2H^+(aq) + 2e^- \rightarrow H_2(g)$	0.00	1
$Br_2(1) + 2e^- \rightarrow 2Br^-(aq)$	+1.07	2
$Ni^{2+}(aq) + 2e^- \rightarrow Ni(s)$	-0.25	3
$Cl_2(g) + 2e^- \rightarrow 2Cl^-(aq)$	+1.36	4
$Fe^{2+}(aq) + 2e^- \rightarrow Fe(s)$	-0.44	5

10.2 For the reaction: $A \rightarrow B$, the activation energy is 30 kcal; 25 kcal of energy is evolved when one mole of A is converted to B. Taking A to have zero energy, construct an activation energy diagram similar to Figure 10.2.

10.3 Referring to Figure 10.5, construct a table of water vapor pressure at $5°$ intervals between 0 and $50°C$.

10.4 For the reaction: $N_2(g) + 3H_2(g) \rightarrow 2NH_3(g)$, the following data are obtained for the variation of the free energy change, ΔG, with temperature, T.

$$\Delta G \text{ (kcal)} \quad -8.0 \quad -3.3 \quad +1.3 \quad +6.0 \quad +10.7$$

$$T \text{ (}°K\text{)} \quad\quad 300 \quad 400 \quad 500 \quad 600 \quad\quad 700$$

Plot the data and use the graph to estimate:
   a. ΔG at $350°K$ b. ΔG at $200°K$ c. the temperature at which $\Delta G = 0$

10.5 For the reaction: $2SO_2(g) + O_2(g) \rightarrow 2SO_3(g)$, ΔG (kcal) $= -47.0 + 0.0453\, T$. Plot ΔG vs. T from 300 to $1000°K$.

10.6 Given that $\Delta G = \Delta H - T\Delta S$, where both ΔH and ΔS are constants, estimate ΔH and ΔS from the data in Problem 10.4.

10.7 The following data were obtained for the concentrations of Ag^+ and CrO_4^{2-} in a solution saturated with Ag_2CrO_4:

$$\text{conc. } CrO_4^{2-} \quad 1.0 \times 10^{-4} \quad 0.25 \times 10^{-4} \quad 0.11 \times 10^{-4} \quad 0.067 \times 10^{-4}$$

$$\text{conc. } Ag^+ \quad\quad 1.0 \times 10^{-4} \quad 2.0 \times 10^{-4} \quad 3.0 \times 10^{-4} \quad 4.0 \quad\quad \times 10^{-4}$$

   a. Plot the concentration of $CrO_4^{2-}(y)$ vs. the concentration of $Ag^+(x)$.
   b. Plot the concentration of CrO_4^{2-} vs. $1/(\text{the concentration of } Ag^+)^2$.
   c. Write the equation relating the concentration of CrO_4^{2-} to the concentration of Ag^+.

10.8 Given the ideal gas law: $PV = nRT$, plot for one mole of an ideal gas:
   a. P vs. T, from 200 to $400°K$ at $V = 10$ liters ($R = 0.0821$ liter atm/mole $°K$).
   b. PV vs. P from 1 to 7 atm at $T = 300°K$.
   c. PV vs. T, from 0 to $1200°K$.

10.9 Given the following data for specific heats:

Element	Atomic Wt.	Specific Heat (cal/g $°C$)
Mg	24.3	0.246
Fe	55.9	0.106
Zn	65.4	0.0925
Ag	108	0.0558

plot specific heat vs. atomic weight and draw a smooth curve through the points.

a. Which of the types of graphs in Figure 10.13 does this most closely resemble?
b. After an appropriate change in variables, make a linear plot of this data.
c. From the plot in (b), estimate the specific heat of calcium, atomic weight = 40.0.

10.10 For the radioactive decomposition of radon, the following data are obtained for the amount of radon (A) remaining as a function of time, t.

A (grams)	10.0	8.2	7.0	5.8	5.0	4.2
t (days)	0	1	2	3	4	5

It is believed that the data fit the equation: $\log A = at + b$. By plotting $\log A$ vs. t, obtain values for the constants a and b;

10.11 The following data were obtained for the dependence of the vapor pressure of chloroform on temperature:

P	61.0	100.5	159.6	246.0
T	273	283	293	303

Use a graphical method to calculate the heat of vaporization, using the equation:

$$\log P = \frac{-\Delta H_{vap}}{(2.30)(1.99)\, T} + \text{constant}$$

10.12 For a water solution of acetic acid, the following equation applies:

$$pH = a \log (\text{conc. HA}) - \frac{1}{2}\log K_a$$

Using the following data:

pH	2.4	2.9	3.4	3.8
conc. HA	1.00	0.100	0.0100	0.00100

Calculate K_a by plotting pH vs. log (concentration of HA) and determining the intercept.

10.13 For a certain reaction, the following data were obtained for the variation of the concentration of reactant with time:

conc.	1.00	0.80	0.63	0.50	0.40	0.32
t(min)	0	1	2	3	4	5

This reaction may be:

"Zero order," in which case $C_0 - C = kt$

"First order," in which case $\log \dfrac{C_0}{C} = \dfrac{kt}{2.3}$

"Second order," in which case $\dfrac{1}{C} - \dfrac{1}{C_0} = kt$

where C_0 is the original concentration, C is the concentration at time t, and k is the rate constant. Using a graphical method, determine the order of the reaction and calculate k.

CHAPTER 11

ERROR ANALYSIS

We saw in Chapter 7 that the errors associated with experimental measurements or with quantities calculated from experimental data can be expressed quite simply in terms of significant figures. The rules governing the use of significant figures amount to an elementary form of error analysis, which is sufficient for most of our purposes in general chemistry. Ordinarily, the quantities that we measure in the general chemistry laboratory are based upon a single trial or, at most, upon duplicate determinations. Such experiments hardly justify a more sophisticated type of error analysis than that described in Chapter 7.

We may, however, have the opportunity to carry out more exact experiments where it is important that we be able to estimate quite accurately the uncertainties associated with individual measurements or with results derived from several successive measurements. Let us suppose, for example, that we are asked to determine the percentage of chlorine in a sample of sodium chloride, given the equipment necessary to determine the percentage to four significant figures and the time to make five successive determinations. We might obtain the data listed in Table 11.1.

There are several questions we might raise concerning this data.

1. Since we were able, in principle, to obtain the percentage of chlorine to ±0.01 per cent, why didn't we get the same value in each trial?
2. Should we reject any of these results? (Trial 2 looks a little dubious).
3. What should we report for the percentage of chlorine? The average of all five trials? The average, omitting trial 2? Some other answer?
4. Should we repeat the experiment a few more times to get a "better" value for the percentage of chlorine?

Table 11.1 Percentage of Cl in NaCl

Trial	1	2	3	4	5
Percentage of Cl	60.50	60.41	60.53	60.54	60.52

5. How confident can we be of the value that we report for the percentage of chlorine? What are the chances that it will be within ±0.01 of the true value? ±0.05?

These are typical of the kinds of questions that we will attempt to find answers for in this chapter. Before proceeding further, it will be necessary to define certain terms that are used repeatedly in error analysis.

11.1 ACCURACY AND PRECISION

The accuracy of a measured quantity indicates the extent to which it agrees with what is believed to be the true value. It is described in terms of the error:

$$\text{Error} = \text{Observed value} - \text{True value} \tag{11.1}$$

The smaller the error, the more accurate a measurement is.

The **precision** of a measurement allows us to estimate its reproducibility. It is described by the **deviation**, which is the difference between the observed value and the average value, obtained from a series of measurements:

$$\text{Deviation} = \text{Observed value} - \text{Average value} \tag{11.2}$$

The smaller the deviation, the more precise the measurement is.

To illustrate the distinction between precision and accuracy, consider the data given in Table 11.1 for the percentage of chlorine in sodium chloride. The average value, which is often referred to as the **arithmetic mean**, is readily found to be 60.50 per cent.

$$\text{Arithmetic mean} = \frac{60.50 + 60.41 + 60.53 + 60.54 + 60.52}{5}$$

$$= \frac{302.50}{5} = 60.50$$

The true value for the percentage of chlorine in sodium chloride is, to four significant figures, 60.66 per cent. (This number is based upon the atomic weights of sodium and chlorine, which have been very carefully determined to *five* significant figures.) Consequently, for the data of Table 11.1, Equations 11.1 and 11.2 become:

$$\text{Error} = \text{Observed value} - 60.66$$

$$\text{Deviation} = \text{Observed value} - 60.50$$

These equations can be used to express the precision and accuracy of each measurement of the percentage of chlorine (Table 11.2).

Table 11.2 Precision and Accuracy of Measured Percentages of Chlorine in NaCl

Trial	Observed value	Error	Deviation	Per cent error	Per cent deviation
1	60.50	−0.16	0.00	−0.26	0.00
2	60.41	−0.25	−0.09	−0.41	−0.15
3	60.53	−0.13	+0.03	−0.21	+0.05
4	60.54	−0.12	+0.04	−0.20	+0.07
5	60.52	−0.14	+0.02	−0.23	+0.03

Clearly, the precision of these measurements is better than their accuracy. In every trial, the error is greater in magnitude than the deviation. In general, we can always expect the precision of measurements to surpass their accuracy. Only if the arithmetic mean coincides exactly with the true value will our measurements be as accurate as they are precise.

Frequently, we refer to the per cent error, or the per cent deviation. These quantities are defined as follows:

$$\text{Per cent error} = \frac{\text{Error}}{\text{True value}} \times 100 \qquad (11.3)$$

$$\text{Per cent deviation} = \frac{\text{Deviation}}{\text{Arithmetic mean}} \times 100 \qquad (11.4)$$

The numbers in the last two columns at the right of Table 11.2 were calculated in this manner.

Example 11.1 Four different students measure the boiling point of a certain organic liquid at 760 mm Hg pressure. Their results are:

$$54.9°C, 54.4°C, 54.1°C, \text{ and } 54.2°C$$

The true boiling point, as determined with a more sophisticated apparatus using an ultrapure sample is, to three significant figures, 54.0°C. Determine, for each measurement, the error, deviation, per cent error, and per cent deviation.

Solution The error can be calculated directly from Equation 11.1. In order to use Equation 11.2 to calculate the deviation, we must first obtain the arithmetic mean. We could do this by adding the four numbers and dividing the sum by four. To reduce the amount of arithmetic, we might write:

$$\text{Arithmetic mean} = 54.0 + \frac{0.9 + 0.4 + 0.1 + 0.2}{4}$$

$$= 54.0 + \frac{1.6}{4} = 54.4$$

For the first measurement, we have:

$$\text{Error} = 54.9 - 54.0 = 0.9;$$

$$\text{Per cent error} = \frac{0.9}{54.0} \times 100 = +2\%$$

$$\text{Deviation} = 54.9 - 54.4 = 0.5;$$

$$\text{Per cent deviation} = \frac{0.5}{54.4} \times 100 = +0.9\%$$

Proceeding similarly for each measurement:

Observed value	Error	Per cent error	Deviation	Per cent deviation
54.9	0.9	+2	0.5	+0.9
54.4	0.4	+0.7	0.0	0.0
54.1	0.1	+0.2	-0.3	-0.6
54.2	0.2	+0.4	-0.2	-0.4

EXERCISES

1. Repeat the calculations in Example 11.1 for temperatures expressed in °K (°K = °C + 273.2°).

2. In 1894, Lord Rayleigh determined the mass of nitrogen gas filling a certain container at a known pressure and temperature. The results of successive weighings were as follows:

2.3102 g, 2.3099 g, 2.3101 g, 2.3100 g, 2.3102 g, 2.3101 g,

2.3103 g, 2.3116 g, and 2.3096 g

The nitrogen he used was prepared from air by removing oxygen, water vapor, and carbon dioxide. When he performed similar experiments with chemically pure nitrogen, he obtained a mass of 2.2997 g. Taking this to be the "true" mass, calculate the deviations and errors of each of the masses listed above.

11.2 TYPES OF ERRORS

The most serious errors that students make in the chemistry laboratory (or in life, for that matter) are ones which could either be avoided or corrected for. These are called **determinate errors.** As an example, consider a student who is attempting to analyze a metal oxide by heating it in a stream of hydrogen to form the pure metal.

$$MO(s) + H_2(g) \rightarrow M(s) + H_2O(g)$$

If he spills part of his sample, his result is likely to show a rather large error. This error could be avoided by being more careful; the only way to "correct" for it would be to repeat the experiment.

As an example of a determinate error which could be corrected for quite readily, consider the dilemma of a student who determines the density of benzene at 20°C by weighing samples of the liquid issuing from a 10 ml pipette. He obtains the following masses:

$$8.681 \text{ g}, 8.678 \text{ g}, 8.683 \text{ g}, \text{ and } 8.678 \text{ g}$$

Finding the arithmetic mean of these masses to be 8.680 g, he divides by 10.00 ml and confidently reports a density of 0.8680 g/ml, trusting that his excellent precision insures high accuracy. Unfortunately, the true value is 0.8790 g/ml; the student has made an error of more than 1 per cent. He failed to realize that a 10 ml pipette does not deliver exactly ten ml of liquid. Had he taken the trouble to calibrate his pipette with distilled water, he would have found that it delivered about 9.90 ml. He would then have reported:

$$\text{density benzene} = 8.680 \text{ g}/9.90 \text{ ml} = 0.877 \text{ g/ml}$$

with an error of about 0.2 per cent.

We can readily predict the direction in which a determinate error will shift the calculated value of a property away from the true value, i.e., whether our answer will be "too large" or "too small." In the example just cited, the student used in his calculations a volume that was too large (10.00 ml vs. 9.90 ml). Consequently, since volume appears in the denominator of the density expression:

$$\text{density} = \text{mass/volume}$$

his reported density was too small (0.868 g/ml vs. 0.879 g/ml).

Example 11.2 A student determines the gram equivalent weight of a metal (the weight that combines with eight grams of oxygen) by using hydrogen to reduce a sample of metal oxide to the pure metal. He makes three weighings:

test tube	mass = A
test tube + metal oxide	mass = B
test tube + metal	mass = C

and calculates the gram equivalent weight using the equation:

$$\text{G.E.W.} = 8.000 \text{ g O} \times \frac{\text{mass metal}}{\text{mass oxygen}} = 8.000 \text{ g O} \times \frac{(C - A)}{(B - C)}$$

How will each of the following determinate errors affect the accuracy of his result (i.e., will they make it larger or smaller than the true value)?

a. In weighing the test tube, he records a mass one gram greater than the true mass.
b. After weighing the metal oxide plus test tube, he spills part of the oxide.
c. He fails to convert all of the metal oxide to metal.

Solution

a. Mass A is too large; B and C are presumably correct. From the equation for G.E.W., we see that the calculated mass of metal, $C - A$, will be too small. Hence, the G.E.W., which is directly proportional to $(C - A)$, will be too small.
b. Here, the recorded masses A and B will be correct. The mass recorded for C will be too small, because some of the sample is lost between the second and third weighings. Looking at the equation for G.E.W., we see that $(C - A)$ will be too small, $(B - C)$ will be too large, and the calculated G.E.W. will be too small.
c. Again, A and B are correct; C is too large because not all of the sample is reduced. Following the reasoning of part (b), we deduce that the calculated G.E.W. will be too large.

Determinate errors account for the fact that the accuracy of experiments rarely equals their precision. We find, however, that even when all determinate errors are eliminated, a series of measurements shows a certain amount of scatter, reflected in deviations from the mean. These deviations are due to what are called **indeterminate errors**. The adjective "indeterminate" implies that these errors cannot be corrected for. In other words, we cannot in any rational way adjust our results to compensate for or to eliminate errors of this type.

Indeterminate errors result from inherent imperfections in the instruments and techniques used to carry out measurements. To illustrate this point, consider the student who determines the density of benzene using a carefully calibrated pipette and a high quality analytical balance. In four trials, he might obtain densities of:

$$0.8782 \text{ g/ml}, \quad 0.8794 \text{ g/ml}, \quad 0.8785 \text{ g/ml}, \quad \text{and} \quad 0.8779 \text{ g/ml}$$

The indeterminate errors that are responsible for deviations in this series of experiments could be due to:
a. Slight fluctuations in temperature; the density of a liquid varies with temperature.
b. Difficulty in reading the balance scale to the nearest 0.001 g.
c. Failure to make the benzene level in the pipette coincide exactly with the graduation mark.
d. ? ? ?

Even though indeterminate errors cannot be corrected for, they can be treated statistically to tell us how they are likely to affect the reliability of our measurements. This type of analysis is based on the so-called normal error curve shown in Figure 11.1. This curve tells us the relative frequency of the various deviations that we can expect to find if we make a large number of measurements. The figure allows us to make some general observations about the magnitude of indeterminate errors.

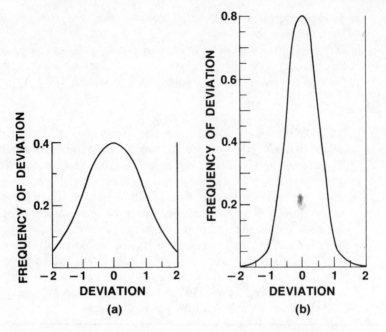

FIGURE 11.1 The error curve: (a), low precision; (b), high precision.

1. Since the curves are symmetric about the midpoint, representing the arithmetic mean, positive and negative deviations are equally likely. This means that we cannot predict the direction in which an indeterminate error will shift a calculated result; values that are "too high" or "too low" are equally likely.

2. Since the curves rise to a maximum at the midpoint, small deviations occur more frequently than large deviations. Indeed, if we observe a very large deviation, we are inclined to suspect that the error may be determinate in nature. For example, in Figure 11.1, a deviation of 10 due to an indeterminate error is extremely unlikely; the chances are that it resulted from a "goof" in the procedure used.

3. The shape of the curve is determined by the inherent precision of the measurement. If the instruments or techniques that we use are relatively crude or "sloppy," we can expect to have an error curve of the type shown at the left of Figure 11.1, with a relatively high frequency of large deviations. As we refine our measurements to improve their precision, we would expect to approach the error distribution shown at the right of Figure 11.1, where large deviations are highly improbable.

EXERCISES

1. A student determining the gram equivalent weight of a metal by reacting it with sulfur makes the following weighings:

                crucible   mass = A

crucible + metal mass = B

crucible + metal sulfide mass = C

$$G.E.W. = 16.0 \text{ g S} \times \frac{\text{mass metal}}{\text{mass sulfur}} = 16.0 \text{ g S} \times \frac{(B - A)}{(C - B)}$$

How will each of the following determinate errors affect the accuracy of his calculated G.E.W.?

a. After weighing the crucible plus metal, he spills some of the metal.
b. Some of the metal sulfide (MS) is oxidized to sulfate (MSO_4).
c. The mass of the crucible plus metal sulfide is incorrectly recorded as 16.200 g; it should have been 16.100 g.

2. Suppose that in 1(c), $A = 15.000$ g, $B = 15.700$ g.
a. What is the true value of the G.E.W.?
b. What value would the student report for the G.E.W.?
c. What is the per cent error in the reported value?

-3. Using Figure 11.1, estimate the relative frequency of obtaining deviations of 0.5 vs. 1.0 for measurements associated with inherently:
a. Low precision (Figure 11.1a).
b. High precision (Figure 11.1b).

4. A class of 20 students measures the length of a certain spectral line with the following results:

2050 Å	2050 Å	2054 Å	2052 Å	2050 Å
2049	2051	2047	2049	2049
2051	2048	2050	2048	2051
2053	2052	2051	2045	2050

a. Find the arithmetic mean and the deviation of each measurement.
b. Construct an "error curve" by plotting the frequency (number) of each deviation vs. the magnitude of the deviation.

11.3 MEASURES OF PRECISION

Two quite different quantities are used to describe the scatter of experimental measurements: the **average deviation** and the **standard deviation**. The average deviation is readily calculated by taking the sum of the deviations, regardless of sign, from the arithmetic mean, and dividing it by the total number of observations.

$$a = \frac{\Sigma |d|}{n} \qquad (11.5)$$

where a is the average deviation, $|d|$ represents the magnitude of an individual deviation, neglecting its sign, and n is the number of trials. (Σ represents the sum.)

Example 11.3 Five students report the following percentages of chlorine in a sample:

$$19.82, 19.57, 19.68, 19.71, \text{ and } 19.75$$

Calculate the arithmetic mean of these results and the average deviation.

Solution To obtain the arithmetic mean, it is convenient to take 19.50 as our base number and write:

$$\text{Arithmetic mean} = 19.50 + \frac{0.32 + 0.07 + 0.18 + 0.21 + 0.25}{5}$$

$$= 19.50 + \frac{1.03}{5} = 19.71$$

(Some people prefer to carry one extra digit in the mean; in this case, they would write 19.706 for the mean. We shall round off to 19.71 to simplify the arithmetic.) We proceed to calculate the individual deviations, d, and the magnitude of each deviation, $|d|$.

Trial	1	2	3	4	5
Percentage of Cl	19.82	19.57	19.68	19.71	19.75
d	+0.11	−0.14	−0.03	0.00	+0.04
$|d|$	0.11	0.14	0.03	0.00	0.04

$$a = \frac{0.11 + 0.14 + 0.03 + 0.00 + 0.04}{5} = 0.06_4$$

The average deviation gives us a qualitative estimate of the precision of our data. Unfortunately, it has no direct statistical significance. The **standard deviation** is a more significant quantity in that it determines the shape of the error curve to be associated with a series of measurements. Unlike the average deviation, it cannot be calculated exactly from a limited amount of experimental data. It can, however, be estimated from the approximate relation:

$$\sigma = \sqrt{\frac{\Sigma d^2}{n-1}} \tag{11.6}$$

where σ is the standard deviation and, as before, d is an individual deviation and n is the number of trials.

Example 11.4 Estimate the standard deviation for the data given in Example 11.3.

Solution We recall that the arithmetic mean is 19.71. Calculating deviations from the mean, we have:

Trial	1	2	3	4	5
Percentage of Cl	19.82	19.57	19.68	19.71	19.75
d	+0.11	−0.14	−0.03	+0.00	+0.04
d^2	0.0121	0.0196	0.0009	0.0000	0.0016

$$\Sigma d^2 = 0.0121 + 0.0196 + 0.0009 + 0.0000 + 0.0016 = 0.0342$$

$$n - 1 = 5 - 1 = 4$$

$$\Sigma d^2 /(n - 1) = 0.0342/4 = 0.0086$$

$$\sigma = (0.0086)^{1/2} = 0.09_2$$

Comparing the answers to Examples 11.3 and 11.4, we see that the standard deviation, 0.092, is greater than the average deviation, 0.064. This discrepancy is ordinarily the case, which may explain why many people prefer to use the average deviation rather than the standard deviation to report the precision of their results! It can be shown that, for a very large number of trials, the standard deviation approaches five-fourths of the average deviation:

$$\sigma \rightarrow \frac{5}{4} a, \text{ as } n \rightarrow \infty \qquad (n = \text{number of trials}) \qquad (11.7)$$

The importance that we attach to the standard deviation reflects the influence it has on the shape of the error curve. It can be shown (see Exercise 3 at the end of this section) that a small value of σ corresponds to a sharp, steeply rising curve, on which the vast majority of deviations are very close to zero. Conversely, a large σ leads to a broad, squat error curve, on which large deviations have a relatively high probability.

In Figure 11.2, we have shown a general form of the error curve similar to those shown in Figure 11.1, except that the divisions along the horizontal axis are expressed as multiples of the standard deviation, σ ($x = -3\sigma, -2\sigma, -\sigma, 0, \sigma, 2\sigma, 3\sigma$). To interpret this curve, let us look at the shaded area, which is bounded on the left and right, respectively, by the vertical lines $x = -\sigma$ and $x = \sigma$. *This area is proportional to the probability of observing a deviation within one unit of σ of the arithmetic* mean, located at the midpoint of the curve. The shaded area comprises a little more than two-thirds (more exactly, 68.4 per cent) of the total area under the error curve. This means that if we were to make a large number of trials, we would expect about two-thirds of them to fall within the range:

$$M - \sigma \text{ to } M + \sigma, \text{ or } M \pm \sigma$$

where M is the arithmetic mean and σ is the standard deviation. Slightly less than one-third of the trials would show larger deviations and hence would fall outside of this range.

As we move farther away from the midpoint of the error curve, let us say to $\pm 2\sigma$, we enclose a still greater portion of the total area. The region between $x = -2\sigma$

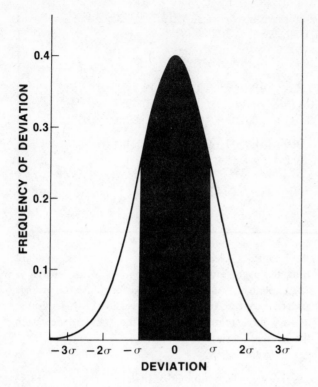

FIGURE 11.2 Normal error curve.

and $x = \mp 2\sigma$ comprises about 95 per cent of the total area. We interpret this to mean that if we made a large number of measurements, we would expect about 95 per cent of them to fall in the range:

$$M - 2\sigma \text{ to } M + 2\sigma, \text{ or } M \pm 2\sigma$$

Only about 5 percent, or 1 in 20, would fall outside this range; i.e., would show a deviation greater in magnitude than 2σ.

A word of caution is in order concerning predictions based on Figure 11.2. Like any predictions made on the basis of the laws of probability, *they can be expected to be valid only when we are dealing with a large number of observations.* With only a few results to work with, we can expect, at best, an error pattern which crudely approximates that shown in Figure 11.2. Indeed, if we have only a few observations, we cannot expect to obtain the true arithmetic mean, M, on which the figure and table are based. It is highly likely that if more results were available, the apparent mean would change to a new value, hopefully approaching more and more closely the true mean of an infinite number of trials.

EXERCISES

1. Calculate a and σ for the data in Table 11.1.
2. A careful experimenter obtains the following values for the atomic weight of cadmium: 112.25, 112.36, 112.32, 112.21, 112.30, and 112.36.

    a. What are the arithmetic mean and the standard deviation?

    b. If the experimenter makes one more measurement, what is the likelihood that it will fall within ±0.06 of the mean?

3. The equation of the error curve (Figure 11.2) is:

$$y = \frac{e^{-x^2/2\sigma^2}}{2.5\sigma}$$

    a. Show that the value of y when $x = 0$ (the midpoint of the curve) is inversely proportional to the standard deviation, σ.

    b. Make an accurate plot of y vs. x in the region $x = -2$ to $x = +2$ for $\sigma = 1$. (Use a table of exponentials from a handbook to evaluate y at various values of x.)

    c. Repeat the plot in (b) for $\sigma = 1/2$, using the same scale of x and y values, and compare the two plots to Figure 11.1.

11.4 RELIABILITY OF THE MEAN

When we carry out a quantitative experiment, we are always restricted to a limited number of trials, often as few as two, seldom more than five or six. The value which we report is ordinarily the arithmetic mean of the several trials. The question arises as to how much confidence we can place in this value. If we could make an infinite number of determinations, how close could we reasonably expect the true mean to come to the mean calculated on the basis of a few trials?

Questions such as this can be answered statistically in terms of what are known as **confidence levels**. Consider, for example, Exercise 2 at the end of Section 11.3, in which we showed that the mean value of the atomic weight of cadmium, based on six trials, was 112.30. It can be shown statistically (see Exercise 1 at the end of this section) that the true mean will fall within ±0.02 of 112.30 at the "50 per cent confidence level," which we interpret to mean that there is a 50-50 chance that the true mean would fall in the range 112.28–112.32. Again, we can show that the true mean will fall within ±0.05 of 112.30 at the "90 per cent confidence level." In other words, the chances are 9 in 10 that the true mean, which could be determined only by making an infinite number of trials, would fall between 112.25 and 112.35.

Reliability limits can be assigned to a calculated mean by using the so-called "t-test," which makes use of the equation:

$$M = \text{Calculated mean} \pm t\sigma/\sqrt{n} \qquad (11.8)$$

where M is the true mean, σ is the standard deviation, n is the number of trials, and t is a statistical parameter which can be obtained from tables such as Table 11.3.

To illustrate how Equation 11.8 is used, let us consider a specific example. Suppose a student, asked to determine the molecular weight of an organic solute, makes three trials with the following results:

159.0, 161.0, and 160.0

Table 11.3   Values of t in Equation 11.8

	Confidence level		
n	*50 per cent*	*90 per cent*	*99 per cent*
2	1.000	6.314	63.657
3	0.816	2.920	9.925
4	0.765	2.353	5.841
5	0.741	2.132	4.604
6	0.727	2.015	4.032
8	0.711	1.895	3.500
10	0.703	1.833	3.250
∞	0.674	1.645	2.576

Clearly, the calculated mean is 160.0. The standard deviation (Equation 11.6) is:

$$\sigma = \sqrt{\frac{\Sigma d^2}{n-1}} = \sqrt{\frac{(1.0)^2 + (1.0)^2}{2}} = 1.0$$

Substituting in Equation 11.8, we obtain:

$$M = 160.0 \pm t(1.0)/\sqrt{3} = 160.0 \pm 1.0t/1.7 = 160.0 \pm 0.59t$$

Reading across Table 11.3 at $n = 3$:

50 per cent level: $M = 160.0 \pm 0.816(0.59)$

                                  $= 160.0 \pm 0.5;$ range = 159.5 - 160.5

90 per cent level: $M = 160.0 \pm 2.920(0.59)$

                                  $= 160.0 \pm 1.7;$ range = 158.3 - 161.7

99 per cent level: $M = 160.0 \pm 9.925(0.59)$

                                  $= 160.0 \pm 5.8;$ range = 154.2 - 165.8

This calculation tells us that if an infinite number of trials were made, there is a 50-50 chance that the true mean would fall within ±0.5 of 160.0, 9 chances in 10 that it would fall within ±1.7 of 160.0, and 99 chances in 100 that it would be within ±5.8. Putting it another way, there is one chance in two that the true mean will fall *outside* the range 159.5-160.5; 1 chance in 10 that it will fall outside the range 158.3-161.7, and only one chance in 100 that it will be less than 154.2 or greater than 165.8.

As this example illustrates, the higher the confidence level we demand, the less precisely we can specify the value of the true mean. To choose two extreme cases: we can be 0 per cent confident that the true mean coincides exactly with the calculated mean ($M = 160.0 \pm 0.0$), but we can be 100 per cent confident that the true mean falls between ∞ and 0!

Example 11.5 The observed mean of a series of determinations of the density of a liquid is 1.69 g/ml. Calculate reliability limits at the 90 per cent confidence level if:

    a. $\sigma = 0.10$, $n = 2$.
    b. $\sigma = 0.05$, $n = 2$
    c. $\sigma = 0.05$; $n = 5$

Solution:

    a. From Table 11.3, we find that for $n = 2$ at the 90 per cent confidence level, $t = 6.314$

$$M = 1.69 \pm \frac{6.3(0.10)}{\sqrt{2}} = 1.69 \pm 0.45$$

    b.
$$M = 1.69 \pm \frac{6.3(0.05)}{\sqrt{2}} = 1.69 \pm 0.22$$

    c. $t = 2.132$; $\quad M = 1.69 \pm \dfrac{2.1(0.05)}{\sqrt{5}} = 1.69 \pm 0.05$

We see from Example 11.5 (and from Equation 11.8) that the reliability of the mean depends upon the magnitude of the standard deviation. Large values of σ imply poor precision and lead to a relatively unreliable mean. The reliability of the calculated mean is also a function of the number of observations upon which it is based. The more trials we make, the more confidence we have that the mean we calculate is a good approximation to the true mean. In Example 11.5 we found that when n increased from 2 to 5, the limits of M, at the 90 per cent confidence level, sharpened from ±0.22 to ±0.05.

It should be noted, however, that beyond a certain point, there is little to be gained by increasing the number of observations. Referring again to Example 11.5, let us see what happens if we make $n = 6$ when $\sigma = 0.05$. From Table 11.3, we find $t = 2.015$ at the 90 per cent level. Hence:

$$M = 1.69 \pm \frac{2.0(0.05)}{\sqrt{6}} = 1.69 \pm 0.04$$

Comparing this result with that calculated in Example 11.5, part (c), we see that increasing n from 5 to 6 had very little effect on the reliability limits of M (±0.04 instead of ±0.05).

Strictly speaking, Equation 11.8 tells us only the deviation which we may reasonably expect between the observed and true mean. *If* we can assume that there are no determinate errors involved, the true mean should coincide with the true value of the quantity we are measuring. In this case, Equation 11.8 should tell us the expected *error* to be associated with the mean. Thus, in Example 11.5, part (c), if the precision and accuracy of the density measurements are the same, we could report the density to be 1.69 ± 0.05 with 90 per cent confidence that the true density would lie within the indicated limits.

EXERCISES

1. Consider the data of Exercise (2), Section 11.3, where $\sigma = 0.060$. What are the reliability limits of the mean at the 50 per cent, 90 per cent, and 99 per cent confidence levels?
2. Repeat the calculations in Example 11.5 at the 99 per cent confidence level.
3. A student, in two trials, finds the values of 16.6 and 17.2 for the percentage of chlorine in a sample. If he wishes to be 99 per cent sure of his answer, what limits should he put on the value he reports?
4. Consider Equation 11.8 with $\sigma = 1$. What are the reliability limits of the mean at the 90 per cent confidence level if n is:
   a. 2   b. 3   c. 4   d. 5   e. 6
   This calculation illustrates the "law of diminishing returns"; the greater the number of measurements that you make, the less justification there is for making another one!

11.5 REJECTION OF A RESULT

Let us suppose that a student in the general chemistry laboratory, asked to find the molecular weight of an organic liquid, makes five determinations with the following results:

$$154.2, 148.0, 153.5, 152.9, \text{ and } 154.5$$

So far as he knows, each run was carried out in the same manner. There was no obvious determinate error in any of the trials. Yet, the number 148.0 appears to be out of line; should it be rejected in calculating the mean?

The question we are really asking is, "Does the value 148.0 reflect a determinate error of which the student was unaware?" If this is the case, it should be rejected. To answer this question, we must ask another. What are the chances that a deviation of this magnitude would turn up normally in carrying out a series of experiments? Unfortunately, there are no hard and fast rules which will give us definitive answers to these questions. Chemists and other scientists frequently use one or another of several empirical rules to decide when to reject a doubtful result.

The 2.5a and 4a rules. These rules are based upon the shape of the error curve (Figures 11.1 and 11.2), which suggests that unusually large deviations are unlikely to result from indeterminate (random) errors. It is assumed that, beyond a certain limit, any such deviations must be due to a determinate error of which the experimenter was unaware. If this is the case, the measurement should of course be rejected.

The 2.5a rule advises us to reject a value if its deviation from the trial mean, calculated by ignoring the doubtful value, is greater than 2.5 times the average deviation, a. Applying this rule to the molecular weight data discussed above, we first take the mean of the four "reliable" values:

$$\frac{154.2 + 153.5 + 152.9 + 154.5}{4} = 153.8$$

$$a = \frac{\Sigma|d|}{n} = \frac{0.4 + 0.3 + 0.9 + 0.7}{4} = 0.6$$

The deviation of the suspected value, 148.0, from the mean, 153.8, is 5.8. Since:

$$5.8 > 2.5(0.6) = 1.5$$

this rule would clearly reject the value 148.0.

The $4a$ rule is similar to the $2.5a$ rule, except that a value is rejected if its deviation from the trial mean exceeds $4a$ rather than $2.5a$. Clearly, if we use this rule, we are less likely to reject doubtful values. If the deviation of the suspected value were 3 times the average deviation, it would be retained if we used the $4a$ rule and rejected if we used the $2.5a$ rule.

In the case of the molecular weight data, the $4a$ rule would advise us to reject the number 148.0.

$$5.8 > 4(0.6) = 2.4$$

A weakness of both of these rules is that they relate directly to the shape of the error curve, which is valid only for a large number of observations. With only a few measurements, we are not in a position to estimate accurately the true mean and hence the true deviation of the suspected value. In the case of the molecular weight determinations, for example, it is quite possible that if we were to carry out ten determinations rather than five, we might obtain the following results:

154.2, 148.0, 153.5, 152.9, 154.5

149.6, 152.0, 148.4, 154.8, 151.2

in which case the trial mean would shift from 153.8 to 152.3, and one can readily show (see Exercise 1 at the end of this section) that both of these rules would advise us to *retain* the number 148.0.

The Q Test. This test, which is somewhat more valid statistically than the $2.5a$ or $4a$ rules, is based on the following procedure.

a. Arrange the results in ascending order:

Example: 148.0, 152.9, 153.5, 154.2, 154.5

b. Obtain the difference between the suspected value and the value nearest to it:

152.9 − 148.0 = 4.9

c. Calculate the range, i.e., the difference between the highest and lowest value, including the suspect:

$$154.5 - 148.0 = 6.5$$

d. Obtain a quotient, Q, by dividing the answer obtained in step (b) by the answer obtained in (c):

$$Q = 4.9/6.5 = 0.75$$

e. Compare to the value of $Q_{0.90}$ found in Table 11.4, which tells us the maximum value that we can expect Q to have, at the 90 per cent confidence level, as the result of a random error. If the Q calculated in (d) exceeds the value of $Q_{0.90}$ found in the table, we reject the suspected value. Looking at Table 11.4, we find that $Q_{0.90}$ for 5 observations ($n = 5$) is 0.64. Since 0.75 is greater than 0.64, we would reject the number 148.0. In principle at least, we do this with 90 per cent confidence, i.e., the chances are less than 1 in 10 that the measurement is valid.

Table 11.4 The Q Test

n	3	4	5	6	7	8	9	10
$Q_{0.90}$	0.94	0.76	0.64	0.56	0.51	0.47	0.44	0.41

This test tends to be more stringent than either of the rules discussed previously, in that it is less likely to allow us to reject a suspected value (Example 11.6).

Example 11.6 A student obtains the following results for the molarity of an NaOH solution: 0.504, 0.510, 0.514, and 0.530. Apply the $4a$ rule and the Q test to decide whether the number 0.530 should be rejected.

Solution

$$4a \text{ rule: trial mean} = \frac{0.504 + 0.510 + 0.514}{3} = 0.509$$

$$a = \frac{\Sigma |d|}{n} = \frac{0.005 + 0.001 + 0.005}{3} = 0.004$$

Deviation of suspected value $= 0.530 - 0.509 = 0.021$

$$0.021 > 4(0.004) = 0.016 \quad \textit{Reject}$$

Q test:

$$\overset{\overset{\displaystyle 0.016}{\longleftrightarrow}}{\underset{\underset{\displaystyle 0.026}{\longleftrightarrow}}{0.504 \quad\quad 0.510 \quad\quad 0.514 \quad\quad 0.530}}$$

$Q = 0.016/0.026 = 0.62$

From Table 11.4, we find $Q_{0.90} = 0.76$ for $n = 4$.
Since $Q < Q_{0.90}$, we should *retain* the value 0.530.

Not infrequently, we find that, with a limited number of trials, the Q test advises us to retain a value which the $2.5a$ rule or the $4a$ rule would reject. What do we do in such a case? The obvious answer is to carry out more determinations, in which case the three rules become more reliable and more nearly consistent. If we cannot do this, we have to accept the fact that statistical considerations are not very helpful when we are limited to a few trials. We may well find ourselves with the dilemma of choosing between the Q test, which may retain invalid results, and the $4a$ and $2.5a$ rules, which are successively more likely to reject valid results.

EXERCISES

1. Apply the $2.5a$ rule and the $4a$ rule to the ten molecular weight values given on p. 189 to see whether 148.0 should be rejected.
2. A student determines the percentage of iron in an ore, obtaining six values:

$$47.0, 55.2, 49.4, 50.1, 49.6, \text{ and } 50.5$$

Apply the Q test successively to the highest and lowest numbers to see if any of them should be rejected. That is, test 55.2 first, then 47.0, then, if necessary, 50.5, and so on.

11.6 ERROR OF A CALCULATED RESULT

The indeterminate error in an individual measurement can be estimated by repeating the measurement several times and seeing how well the several values check one another. For example, after using a particular balance to make repeated weighings, we might estimate that the uncertainty associated with each weighing is about ± 0.002 g. Again, from experience gained with a particular 10-ml pipette, we might conclude that it can be depended upon to deliver 10.04 ± 0.01 ml of liquid.

However, most of the quantities that we determine in the laboratory are based upon more than one measured quantity. We might, for example, determine the density of a liquid using an analytical balance with a reproducibility of ± 0.002 g. and a pipette which delivers to ± 0.01 ml. How should these errors be combined to estimate the resultant error in the density?

In Chapter 7, we described how questions such as this could be answered in an approximate way by using the rules of significant figures. A more exact treatment leads to the rules given in Table 11.5.

Table 11.5   Indeterminate Error in a Derived Result

Operation	Example	Error
1. Addition	$S = A + B$	$E_S^2 = E_A^2 + E_B'^2$
2. Subtraction	$D = A - B$	$E_D^2 = E_A^2 + E_B^2$
3. Multiplication	$P = A \times B$	$\left(\dfrac{E_P}{P}\right)^2 = \left(\dfrac{E_A}{A}\right)^2 + \left(\dfrac{E_B}{B}\right)^2$
4. Division	$Q = A/B$	$\left(\dfrac{E_Q}{Q}\right)^2 = \left(\dfrac{E_A}{A}\right)^2 + \left(\dfrac{E_B}{B}\right)^2$

The way in which Table 11.5 is used is illustrated in Examples 11.7 and 11.8.

Example 11.7   A chemist prepares a compound containing only carbon, hydrogen, and oxygen. He sends it out for C-H analysis with the following results:

$$\% \, C = 54.80 \pm 0.05; \quad \% \, H = 6.92 \pm 0.05$$

What is the uncertainty in the per cent of oxygen, calculated by difference from 100?

Solution   The calculation is:

$$\% \, O = 100.00 - 54.80 - 6.92 = 38.28$$

Applying the rule for subtraction in Table 11.5:

$$E_O^2 = E_C^2 + E_H^2 = (0.05)^2 + (0.05)^2 = 50 \times 10^{-4}$$
$$E_O = (50 \times 10^{-4})^{1/2} = 7 \times 10^{-2} = 0.07$$

In other words: $\% \, O = 38.38 \pm 0.07$

Example 11.8   To determine the density of a liquid, a student allows a pipette to drain into a weighed Erlenmeyer flask and then reweighs. His data, along with the estimated uncertainties, are:

$$\text{volume liquid} = 10.04 \pm 0.01 \text{ ml}$$
$$\text{mass empty flask} = 22.452 \pm 0.002 \text{ g}$$
$$\text{mass (flask + liquid)} = 33.629 + 0.002 \text{ g}$$

He calculates the density to be:

$$d = \frac{m}{v} = \frac{33.629 \text{ g} - 22.452 \text{ g}}{10.04 \text{ ml}} = \frac{11.177 \text{ g}}{10.04 \text{ ml}} = 1.113 \text{ g/ml}$$

Estimate the error in:
    a. The mass of liquid.
    b. The density.

Solution

    a. Using the rule for addition:

$$E_m{}^2 = (2 \times 10^{-3})^2 + (2 \times 10^{-3})^2 = 8 \times 10^{-6}$$
$$E_m = (8 \times 10^{-6})^{1/2} = 2.8 \times 10^{-3} \text{ g}$$

    b. Using the rule for division:

$$\left(\frac{E_d}{d}\right)^2 = \left(\frac{E_m}{m}\right)^2 + \left(\frac{E_v}{v}\right)^2 = \left(\frac{2.8 \times 10^{-3}}{11.177}\right)^2 + \left(\frac{1 \times 10^{-2}}{10.04}\right)^2$$

$$= (0.25 \times 10^{-3})^2 + (1.0 \times 10^{-3})^2 = 1.1 \times 10^{-6}$$

$$\frac{E_d}{d} \approx 1 \times 10^{-3}; \quad E_d = (1.113 \text{ g/ml})(1 \times 10^{-3}) = 1 \times 10^{-3} \text{ g/ml}$$

Thus, the density should be reported as 1.113 ± 0.001 g/ml

You will note from Example 11.8 that the error in the density is due almost entirely to the error in the volume. In the calculation, the quantity $(E_m/m)^2$ virtually disappears, so:

$$\left(\frac{E_d}{d}\right)^2 \approx \left(\frac{E_v}{v}\right)^2; \quad \frac{E_d}{d} \approx \frac{E_v}{v}$$

It is generally true that, where indeterminate errors are involved, it is the error in the least accurate quantity that determines the magnitude of the overall error. This is the justification for the empirical rule stated in Chapter 7, that in multiplication or division, the number of significant figures retained in the answer should be that in the least accurate quantity entering the calculations.

EXERCISE
1. A. student determines the gram equivalent weight of iron by reducing a weighed sample of an oxide of iron to the metal and using the equation:

$$\text{G.E.W. Fe} = 8.000 \text{ g} \times \frac{\text{wt. iron}}{\text{wt. oxygen}}$$

His weighings along with the estimated indeterminate errors are

$$\text{empty test tube} = 12.602 \pm 0.002 \text{ g}$$

$$\text{test tube} + \text{iron oxide} = 14.709 \pm 0.002 \text{ g}$$

$$\text{test tube} + \text{iron} = 14.076 \pm 0.002 \text{ g}$$

Estimate the error in:

a. The weight of iron. b. The weight of oxygen. c. The calculated G.E.W.

PROBLEMS

11.1 A student determines the percentage of chlorine in a compound by titrating a weighed sample with $AgNO_3$. His measurements are:

mass beaker	A		volume $AgNO_3$	V
mass beaker + sample	B		conc. $AgNO_3$	M

$$\text{percentage of Cl} = \frac{100 \times V \times M \times 35.5}{(B - A)}$$

What effect will each of the following determinate errors have on the value he reports for the percentage of chlorine?

a. He "overshoots" the end point of the titration, adding excess $AgNO_3$.
b. The measured mass of the beaker is one gram greater than the true value.
c. The silver nitrate solution is diluted with water before titrating.
d. Before titration, the sample is dissolved in water containing Cl^- ions.

11.2 A student analyzing a sample for the percentage of bromine makes four trials with the following results:

$$36.0, 36.3, 35.8, \text{ and } 36.3$$

Calculate:

a. The arithmetic mean.
b. The deviation and per cent deviation of each trial.
c. The average deviation.
d. The standard deviation.

11.3 The atomic weight of osmium was reported to be 190.2, based on the mean of 6 determinations with a standard deviation of 0.1. Using Table 11.3, calculate the reliability of the mean at the 50 per cent, 90 per cent, and 99 per cent confidence levels.

11.4 A class of 10 students reports the following results for the percentage of chlorine in a sample:

$$20.30, 20.18, 19.98, 20.06, 20.22,$$

$$20.08, 20.02, 20.14, 20.07, \text{ and } 20.00$$

Calculate the arithmetic mean and the reliability limits for the mean at the 90 per cent confidence level. What, precisely, do these limits mean? Under what conditions could we say, with 90 per cent confidence, that the true percentage of chlorine lies within these limits?

11.5 A student uses four different voltmeters to measure the voltage of a zinc-copper cell. He obtains the following results:

$$1.10, 1.21, 1.08, \text{ and } 1.12$$

Apply the Q test to decide whether any of these measurements should be neglected, indicating a faulty voltmeter.

11.6 A class of 10 students determines the percentage of aluminum in an Al-Zn alloy. The results are as follows:

$$60, 62, 39, 64, 58, 56, 61, 60, 69, \text{ and } 62$$

The instructor wishes to establish a reliable mean from these results. Apply the Q test to decide which, if any, of the results should be rejected.

11.7 For a certain reaction, the estimated error in ΔH is ± 0.1 kcal. The estimated error in ΔG, measured at $300°K$, is also ± 0.1 kcal. Using the equation:

$$\Delta G = \Delta H - T\Delta S$$

in both calculations:
   a. Estimate the error in ΔS.
   b. Estimate the error in the calculated value of ΔG at $600°K$.

11.8 The molecular weight of a gas can be calculated from the equation:

$$M = \frac{g \times R \times T}{P \times V}$$

where g = mass of gas, R = gas constant = 82.1 ml atm/mole $°K$, T = temperature in $°K$, P = pressure in atmospheres = pressure in mm Hg/760, V = volume in ml.
   A student obtains the following values, with the indicated indeterminate errors:

$$T = 298.5 \pm 0.1°K$$

$$P = 715 \pm 1 \text{ mm Hg}$$

$$V = 260 \pm 1 \text{ ml}$$

$$g = 1.545 \text{ g} \pm 1 \text{ mg}$$

Calculate the molecular weight of the gas and estimate the error.

APPENDIX 1

REFERENCES

General

Richardson, M.: Fundamentals of Mathematics, 3rd ed. Macmillan, 1966.
May, K. O.: Elementary Analysis. John Wiley and Sons, Inc., 1952.
Beiser, A.: Essential Math for the Sciences. McGraw-Hill, 1969.
Marion, J. B., and Davidson, R. C.: Mathematical Preparation for General Physics. W. B. Saunders Company, 1972.
Swartz, C.: Used Math for the First Two Years of College Science. Prentice-Hall, 1973.

Unit Conversions (Chapter 2)

Benson, S.: Chemical Calculations, 3rd ed. John Wiley and Sons, Inc., 1971.

The Slide Rule (Chapter 6)

Saffold, R., and Smalley, A.: The Slide Rule. Doubleday and Co., Inc., 1962.
Breckinridge, W. E.: The Polyphase Slide Rule. Keuffel and Esser, 1944.

Functional Relationships. Graphs (Chapters 9, 10)

Ellerby, G.: Graphs and Calculus. Pergamon Press, 1964.
Davis, D. S.: Empirical Equations and Nomography. McGraw-Hill, 1943.
Johnson, M.: Problem Solving and Chemical Calculations. Harcourt, Brace and World, 1969 (Chapter 12).

Error Analysis (Chapter 11)

Christian, G.: Analytical Chemistry. Xerox College Publishing, 1971 (Chapter 25).
Laitinen, H. A.: Chemical Analysis. McGraw-Hill, 1960 (Chapter 26).
Pierce, C., and Smith, R. N.: General Chemistry Workbook. W. H. Freeman and Co., 1971.
Mills, F. C.: Introduction to Statistics. Holt, Rinehart and Winston, 1956.

APPENDIX 2

MATHEMATICAL TABLES AND SYMBOLS

A. LOGARITHMS

	0	1	2	3	4	5	6	7	8	9
1.0	.0000	.0043	.0086	.0128	.0170	.0212	.0253	.0294	.0334	.0374
1.1	.0414	.0453	.0492	.0531	.0569	.0607	.0645	.0682	.0719	.0755
1.2	.0792	.0828	.0864	.0899	.0934	.0969	.1004	.1038	.1072	.1106
1.3	.1139	.1173	.1206	.1239	.1271	.1303	.1335	.1367	.1399	.1430
1.4	.1461	.1492	.1523	.1553	.1584	.1614	.1644	.1673	.1703	.1732
1.5	.1761	.1790	.1818	.1847	.1875	.1903	.1931	.1959	.1987	.2014
1.6	.2041	.2068	.2095	.2122	.2148	.2175	.2201	.2227	.2253	.2279
1.7	.2304	.2330	.2355	.2380	.2405	.2430	.2455	.2480	.2504	.2529
1.8	.2553	.2577	.2601	.2625	.2648	.2672	.2695	.2718	.2742	.2765
1.9	.2788	.2810	.2833	.2856	.2878	.2900	.2923	.2945	.2967	.2989
2.0	.3010	.3032	.3054	.3075	.3096	.3118	.3139	.3160	.3181	.3201
2.1	.3222	.3243	.3263	.3284	.3304	.3324	.3345	.3365	.3385	.3404
2.2	.3424	.3444	.3464	.3483	.3502	.3522	.3541	.3560	.3579	.3598
2.3	.3617	.3636	.3655	.3674	.3692	.3711	.3729	.3747	.3766	.3784
2.4	.3802	.3820	.3838	.3856	.3874	.3892	.3909	.3927	.3945	.3962
2.5	.3979	.3997	.4014	.4031	.4048	.4065	.4082	.4099	.4116	.4133
2.6	.4150	.4166	.4183	.4200	.4216	.4232	.4249	.4265	.4281	.4298
2.7	.4314	.4330	.4346	.4362	.4378	.4393	.4409	.4425	.4440	.4456
2.8	.4472	.4487	.4502	.4518	.4533	.4548	.4564	.4579	.4594	.4609
2.9	.4624	.4639	.4654	.4669	.4683	.4698	.4713	.4728	.4742	.4757
3.0	.4771	.4786	.4800	.4814	.4829	.4843	.4857	.4871	.4886	.4900
3.1	.4914	.4928	.4942	.4955	.4969	.4983	.4997	.5011	.5024	.5038
3.2	.5051	.5065	.5079	.5092	.5105	.5119	.5132	.5145	.5159	.5172
3.3	.5185	.5198	.5211	.5224	.5237	.5250	.5263	.5276	.5289	.5302
3.4	.5315	.5328	.5340	.5353	.5366	.5378	.5391	.5403	.5416	.5428
3.5	.5441	.5453	.5465	.5478	.5490	.5502	.5514	.5527	.5539	.5551
3.6	.5563	.5575	.5587	.5599	.5611	.5623	.5635	.5647	.5658	.5670
3.7	.5682	.5694	.5705	.5717	.5729	.5740	.5752	.5763	.5775	.5786
3.8	.5798	.5809	.5821	.5832	.5843	.5855	.5866	.5877	.5888	.5899
3.9	.5911	.5922	.5933	.5944	.5955	.5966	.5977	.5988	.5999	.6010
4.0	.6021	.6031	.6042	.6053	.6064	.6075	.6085	.6096	.6107	.6117
4.1	.6128	.6138	.6149	.6160	.6170	.6180	.6191	.6201	.6212	.6222
4.2	.6232	.6243	.6253	.6263	.6274	.6284	.6294	.6304	.6314	.6325
4.3	.6335	.6345	.6355	.6365	.6375	.6385	.6395	.6405	.6415	.6425
4.4	.6435	.6444	.6454	.6464	.6474	.6484	.6493	.6503	.6513	.6522
4.5	.6532	.6542	.6551	.6561	.6571	.6580	.6590	.6599	.6609	.6618
4.6	.6628	.6637	.6646	.6656	.6665	.6675	.6684	.6693	.6702	.6712
4.7	.6721	.6730	.6739	.6749	.6758	.6767	.6776	.6785	.6794	.6803
4.8	.6812	.6821	.6830	.6839	.6848	.6857	.6866	.6875	.6884	.6893
4.9	.6902	.6911	.6920	.6928	.6937	.6946	.6955	.6964	.6972	.6981

A. LOGARITHMS (Continued)

	0	1	2	3	4	5	6	7	8	9
5.0	.6990	.6998	.7007	.7016	.7024	.7033	.7042	.7050	.7059	.7067
5.1	.7076	.7084	.7093	.7101	.7110	.7118	.7126	.7135	.7143	.7152
5.2	.7160	.7168	.7177	.7185	.7193	.7202	.7210	.7218	.7226	.7235
5.3	.7243	.7251	.7259	.7267	.7275	.7284	.7292	.7300	.7308	.7316
5.4	.7324	.7332	.7340	.7348	.7356	.7364	.7372	.7380	.7388	.7396
5.5	.7404	.7412	.7419	.7427	.7435	.7443	.7451	.7459	.7466	.7474
5.6	.7482	.7490	.7497	.7505	.7513	.7520	.7528	.7536	.7543	.7551
5.7	.7559	.7566	.7574	.7582	.7589	.7597	.7604	.7612	.7619	.7627
5.8	.7634	.7642	.7649	.7657	.7664	.7672	.7679	.7686	.7694	.7701
5.9	.7709	.7716	.7723	.7731	.7738	.7745	.7752	.7760	.7767	.7774
6.0	.7782	.7789	.7796	.7803	.7810	.7818	.7825	.7832	.7839	.7846
6.1	.7853	.7860	.7868	.7875	.7882	.7889	.7896	.7903	.7910	.7917
6.2	.7924	.7931	.7938	.7945	.7952	.7959	.7966	.7973	.7980	.7987
6.3	.7993	.8000	.8007	.8014	.8021	.8028	.8035	.8041	.8048	.8055
6.4	.8062	.8069	.8075	.8082	.8089	.8096	.8102	.8109	.8116	.8122
6.5	.8129	.8136	.8142	.8149	.8156	.8162	.8169	.8176	.8182	.8189
6.6	.8195	.8202	.8209	.8215	.8222	.8228	.8235	.8241	.8248	.8254
6.7	.8261	.8267	.8274	.8280	.8287	.8293	.8299	.8306	.8312	.8319
6.8	.8325	.8331	.8338	.8344	.8351	.8357	.8363	.8370	.8376	.8382
6.9	.8388	.8395	.8401	.8407	.8414	.8420	.8426	.8432	.8439	.8445
7.0	.8451	.8457	.8463	.8470	.8476	.8482	.8488	.8494	.8500	.8506
7.1	.8513	.8519	.8525	.8531	.8537	.8543	.8549	.8555	.8561	.8567
7.2	.8573	.8579	.8585	.8591	.8597	.8603	.8609	.8615	.8621	.8627
7.3	.8633	.8639	.8645	.8651	.8657	.8663	.8669	.8675	.8681	.8686
7.4	.8692	.8698	.8704	.8710	.8716	.8722	.8727	.8733	.8739	.8745
7.5	.8751	.8756	.8762	.8768	.8774	.8779	.8785	.8791	.8797	.8802
7.6	.8808	.8814	.8820	.8825	.8831	.8837	.8842	.8848	.8854	.8859
7.7	.8865	.8871	.8876	.8882	.8887	.8893	.8899	.8904	.8910	.8915
7.8	.8921	.8927	.8932	.8938	.8943	.8949	.8954	.8960	.8965	.8971
7.9	.8976	.8982	.8987	.8993	.8998	.9004	.9009	.9015	.9020	.9026
8.0	.9031	.9036	.9042	.9047	.9053	.9058	.9063	.9069	.9074	.9079
8.1	.9085	.9090	.9096	.9101	.9106	.9112	.9117	.9122	.9128	.9133
8.2	.9138	.9143	.9149	.9154	.9159	.9165	.9170	.9175	.9180	.9186
8.3	.9191	.9196	.9201	.9206	.9212	.9217	.9222	.9227	.9232	.9238
8.4	.9243	.9248	.9253	.9258	.9263	.9269	.9274	.9279	.9284	.9289
8.5	.9294	.9299	.9304	.9309	.9315	.9320	.9325	.9330	.9335	.9340
8.6	.9345	.9350	.9355	.9360	.9365	.9370	.9375	.9380	.9385	.9390
8.7	.9395	.9400	.9405	.9410	.9415	.9420	.9425	.9430	.9435	.9440
8.8	.9445	.9450	.9455	.9460	.9465	.9469	.9474	.9479	.9484	.9489
8.9	.9494	.9499	.9504	.9509	.9513	.9518	.9523	.9528	.9533	.9538
9.0	.9542	.9547	.9552	.9557	.9562	.9566	.9571	.9576	.9581	.9586
9.1	.9590	.9595	.9600	.9605	.9609	.9614	.9619	.9624	.9628	.9633
9.2	.9638	.9643	.9647	.9652	.9657	.9661	.9666	.9671	.9675	.9680
9.3	.9685	.9689	.9694	.9699	.9703	.9708	.9713	.9717	.9722	.9727
9.4	.9731	.9736	.9741	.9745	.9750	.9754	.9759	.9763	.9768	.9773
9.5	.9777	.9782	.9786	.9791	.9795	.9800	.9805	.9809	.9814	.9818
9.6	.9823	.9827	.9832	.9836	.9841	.9845	.9850	.9854	.9859	.9863
9.7	.9868	.9872	.9877	.9881	.9886	.9890	.9894	.9899	.9903	.9908
9.8	.9912	.9917	.9921	.9926	.9930	.9934	.9939	.9943	.9948	.9952
9.9	.9956	.9961	.9965	.9969	.9974	.9978	.9983	.9987	.9991	.9996

B. SQUARES, CUBES, AND ROOTS*

n	n^2	$\sqrt{n}$	$\sqrt{10n}$	n^3	$\sqrt[3]{n}$	$\sqrt[3]{10n}$	$\sqrt[3]{100n}$
1	1	1.000 000	3.162 278	1	1.000 000	2.154 435	4.641 589
2	4	1.414 214	4.472 136	8	1.259 921	2.714 418	5.848 035
3	9	1.732 051	5.477 226	27	1.442 250	3.107 233	6.694 330
4	16	2.000 000	6.324 555	64	1.587 401	3.419 952	7.368 063
5	25	2.236 068	7.071 068	125	1.709 976	3.684 031	7.937 005
6	36	2.449 490	7.745 967	216	1.817 121	3.914 868	8.434 327
7	49	2.645 751	8.366 600	343	1.912 931	4.121 285	8.879 040
8	64	2.828 427	8.944 272	512	2.000 000	4.308 869	9.283 178
9	81	3.000 000	9.486 833	729	2.080 084	4.481 405	9.654 894
10	100	3.162 278	10.00000	1 000	2.154 435	4.641 589	10.00000
11	121	3.316 625	10.48809	1 331	2.223 980	4.791 420	10.32280
12	144	3.464 102	10.95445	1 728	2.289 428	4.932 424	10.62659
13	169	3.605 551	11.40175	2 197	2.351 335	5.065 797	10.91393
14	196	3.741 657	11.83216	2 744	2.410 142	5.192 494	11.18689
15	225	3.872 983	12.24745	3 375	2.466 212	5.313 293	11.44714
16	256	4.000 000	12.64911	4 096	2.519 842	5.428 835	11.69607
17	289	4.123 106	13.03840	4 913	2.571 282	5.539 658	11.93483
18	324	4.242 641	13.41641	5 832	2.620 741	5.646 216	12.16440
19	361	4.358 899	13.78405	6 859	2.668 402	5.748 897	12.38562
20	400	4.472 136	14.14214	8 000	2.714 418	5.848 035	12.59921
21	441	4.582 576	14.49138	9 261	2.758 924	5.943 922	12.80579
22	484	4.690 416	14.83240	10 648	2.802 039	6.036 811	13.00591
23	529	4.795 832	15.16575	12 167	2.843 867	6.126 926	13.20006
24	576	4.898 979	15.49193	13 824	2.884 499	6.214 465	13.38866
25	625	5.000 000	15.81139	15 625	2.924 018	6.299 605	13.57209
26	676	5.099 020	16.12452	17 576	2.962 496	6.382 504	13.75069
27	729	5.196 152	16.43168	19 683	3.000 000	6.463 304	13.92477
28	784	5.291 503	16.73320	21 952	3.036 589	6.542 133	14.09460
29	841	5.385 165	17.02939	24 389	3.072 317	6.619 106	14.26043
30	900	5.477 226	17.32051	27 000	3.107 233	6.694 330	14.42250
31	961	5.567 764	17.60682	29 791	3.141 381	6.767 899	14.58100
32	1 024	5.656 854	17.88854	32 768	3.174 802	6.839 904	14.73613
33	1 089	5.744 563	18.16590	35 937	3.207 534	6.910 423	14.88806
34	1 156	5.830 952	18.43909	39 304	3.239 612	6.979 532	15.03695
35	1 225	5.916 080	18.70829	42 875	3.271 066	7.047 299	15.18294
36	1 296	6.000 000	18.97367	46 656	3.301 927	7.113 787	15.32619
37	1 369	6.082 763	19.23538	50 653	3.332 222	7.179 054	15.46680
38	1 444	6.164 414	19.49359	54 872	3.361 975	7.243 156	15.60491
39	1 521	6.244 998	19.74842	59 319	3.391 211	7.306 144	15.74061

*Roots of numbers other than those given in this table may be determined from the following relations.

Square Roots:

$$\sqrt{1000n} = 10\sqrt{10n}; \quad \sqrt{100n} = 10\sqrt{n}; \quad \sqrt{\frac{n}{10}} = \frac{\sqrt{10n}}{10}; \quad \sqrt{\frac{n}{100}} = \frac{\sqrt{n}}{10}; \quad \sqrt{\frac{n}{1000}} = \frac{\sqrt{10n}}{100}$$

Cube Roots:

$$\sqrt[3]{100,000n} = 10\sqrt[3]{100n}; \quad \sqrt[3]{10,000n} = 10\sqrt[3]{10n}; \quad \sqrt[3]{1000n} = 10\sqrt[3]{n}; \quad \sqrt[3]{\frac{n}{100}} = \frac{\sqrt[3]{10n}}{10}; \quad \sqrt[3]{\frac{n}{1000}} = \frac{\sqrt[3]{n}}{10}$$

n	n^2	$\sqrt{n}$	$\sqrt{10n}$	n^3	$\sqrt[3]{n}$	$\sqrt[3]{10n}$	$\sqrt[3]{100n}$
40	1 600	6.324 555	20.00000	64 000	3.419 952	7.368 063	15.87401
41	1 681	6.403 124	20.24846	68 921	3.448 217	7.428 959	16.00521
42	1 764	6.480 741	20.49390	74 088	3.476 027	7.488 872	16.13429
43	1 849	6.557 439	20.73644	79 507	3.503 398	7.547 842	16.26133
44	1 936	6.633 250	20.97618	85 184	3.530 348	7.605 905	16.38643
45	2 025	6.708 204	21.21320	91 125	3.556 893	7.663 094	16.50964
46	2 116	6.782 330	21.44761	97 336	3.583 048	7.719 443	16.63103
47	2 209	6.855 655	21.67948	103 823	3.608 826	7.774 980	16.75069
48	2 304	6.928 203	21.90890	110 592	3.634 241	7.829 735	16.86865
49	2 401	7.000 000	22.13594	117 649	3.659 306	7.883 735	16.98499
50	2 500	7.071 068	22.36068	125 000	3.684 031	7.937 005	17.09976
51	2 601	7.141 428	22.58318	132 651	3.708 430	7.989 570	17.21301
52	2 704	7.211 103	22.80351	140 608	3.732 511	8.041 452	17.32478
53	2 809	7.280 110	23.02173	148 877	3.756 286	8.092 672	17.43513
54	2 916	7.348 469	23.23790	157 464	3.779 763	8.143 253	17.54411
55	3 025	7.416 198	23.45208	166 375	3.802 952	8.193 213	17.65174
56	3 136	7.483 315	23.66432	175 616	3.825 862	8.242 571	17.75808
57	3 249	7.549 834	23.87467	185 193	3.848 501	8.291 344	17.86316
58	3 364	7.615 773	24.08319	195 112	3.870 877	8.339 551	17.96702
59	3 481	7.681 146	24.28992	205 379	3.892 996	8.387 207	18.06969
60	3 600	7.745 967	24.49490	216 000	3.914 868	8.434 327	18.17121
61	3 721	7.810 250	24.69818	226 981	3.936 497	8.480 926	18.27160
62	3 844	7.874 008	24.89980	238 328	3.957 892	8.527 019	18.37091
63	3 969	7.937 254	25.09980	250 047	3.979 057	8.572 619	18.46915
64	4 096	8.000 000	25.29822	262 144	4.000 000	8.617 739	18.56636
65	4 225	8.062 258	25.49510	274 625	4.020 726	8.662 391	18.66256
66	4 356	8.124 038	25.69047	287 496	4.041 240	8.706 588	18.75777
67	4 489	8.185 353	25.88436	300 763	4.061 548	8.750 340	18.85204
68	4 624	8.246 211	26.07681	314 432	4.081 655	8.793 659	18.94536
69	4 761	8.306 624	26.26785	328 509	4.101 566	8.836 556	19.03778
70	4 900	8.366 600	26.45751	343 000	4.121 285	8.879 040	19.12931
71	5 041	8.426 150	26.64583	357 911	4.140 818	8.921 121	19.21997
72	5 184	8.485 281	26.83282	373 248	4.160 168	8.962 809	19.30979
73	5 329	8.544 004	27.01851	389 017	4.179 339	9.004 113	19.39877
74	5 476	8.602 325	27.20294	405 224	4.198 336	9.045 042	19.48695
75	5 625	8.660 254	27.38613	421 875	4.217 163	9.085 603	19.57434
76	5 776	8.717 798	27.56810	438 976	4.235 824	9.125 805	19.66095
77	5 929	8.774 964	27.74887	456 533	4.254 321	9.165 656	19.74681
78	6 084	8.831 761	27.92848	474 552	4.272 659	9.205 164	19.83192
79	6 241	8.888 194	28.10694	493 039	4.290 840	9.244 335	19.91632
80	6 400	8.944 272	28.28427	512 000	4.308 869	9.283 178	20.00000
81	6 561	9.000 000	28.46050	531 441	4.326 749	9.321 698	20.08299
82	6 724	9.055 385	28.63564	551 368	4.344 481	9.359 902	20.16530
83	6 889	9.110 434	28.80972	571 787	4.362 071	9.397 796	20.24694
84	7 056	9.165 151	28.98275	592 704	4.379 519	9.435 388	20.32793
85	7 225	9.219 544	29.15476	614 125	4.396 830	9.472 682	20.40828
86	7 396	9.273 618	29.32576	636 056	4.414 005	9.509 685	20.48800
87	7 569	9.327 379	29.49576	658 503	4.431 048	9.546 403	20.56710
88	7 744	9.380 832	29.66479	681 472	4.447 960	9.582 840	20.64560
89	7 921	9.433 981	29.83287	704 969	4.464 745	9.619 002	20.72351
90	8 100	9.486 833	30.00000	729 000	4.481 405	9.654 894	20.80084
91	8 281	9.539 392	30.16621	753 571	4.497 941	9.690 521	20.87759
92	8 464	9.591 663	30.33150	778 688	4.514 357	9.725 888	20.95379

n	n^2	$\sqrt{n}$	$\sqrt{10n}$	n^3	$\sqrt[3]{n}$	$\sqrt[3]{10n}$	$\sqrt[3]{100n}$
93	8 649	9.643 651	30.49590	804 357	4.530 655	9.761 000	21.02944
94	8 836	9.695 360	30.65942	830 584	4.546 836	9.795 861	21.10454
95	9 025	9.746 794	30.82207	857 375	4.562 903	9.830 476	21.17912
96	9 216	9.797 959	30.98387	884 736	4.578 857	9.864 848	21.25317
97	9 409	9.848 858	31.14482	912 673	4.594 701	9.898 983	21.32671
98	9 604	9.899 495	31.30495	941 192	4.610 436	9.932 884	21.39975
99	9 801	9.949 874	31.46427	970 299	4.626 065	9.966 555	21.47229
100	10 000	10.00000	31.62278	1 000 000	4.641 589	10.00000	21.54435

C. EXPONENTIAL FUNCTIONS

x	e^x	e^{-x}	x	e^x	e^{-x}
0	1.000	1.000	1.6	4.953	0.202
0.1	1.105	0.905	1.7	5.474	0.183
0.2	1.221	0.819	1.8	6.050	0.165
0.3	1.350	0.741	1.9	6.686	0.150
0.4	1.492	0.670	2.0	7.389	0.135
0.5	1.649	0.607	2.5	12.18	0.0821
0.6	1.822	0.549	3.0	20.09	0.0498
0.7	2.014	0.497	3.5	33.12	0.0302
0.8	2.226	0.449	4.0	54.60	0.0183
0.9	2.460	0.407	4.5	90.02	0.0111
1.0	2.718	0.368	5.0	148.4	0.00674
1.1	3.004	0.333	6.0	403.4	0.00248
1.2	3.320	0.301	7.0	1097	0.000912
1.3	3.669	0.273	8.0	2981	0.000336
1.4	4.055	0.247	9.0	8103	0.000123
1.5	4.482	0.223	10.0	22026	0.0000454

Between $x = 0$ and $x = 2$, e^x and e^{-x} for values of x not listed directly can be estimated by interpolation. For example, $e^{0.65}$ is approximately half-way between $e^{0.6}$ (1.822) and $e^{0.7}$ (2.014), i.e., $e^{0.65} \approx 1.918$. More extensive tables of exponential functions can be found in various handbooks, e.g., *Handbook of Chemistry and Physics*, Chemical Rubber Publishing Co. Finally, e^x and e^{-x} can be evaluated by taking logarithms, as explained in Chapter 5.

D. SYMBOLS USED IN MATHEMATICS

$+$	plus
$-$	minus
$\pm$	plus or minus
$\times$	multiplied by; may also be indicated by a dot between the factors or by enclosing the factors within parentheses. $6 \times 2 = 6 \cdot 2 = (6)(2) = 12$
$\div$	divided by; also indicated by writing the divisor under the dividend with a line between, or separating by a slash. $6 \div 2 = \dfrac{6}{2} = 6/2 = 3$
$=$	equals
$\equiv$	is identical to; is defined as
$\approx$	is approximately
$\hat{=}$	is equivalent to
$>$	is greater than
$<$	is less than
$\leqslant$	is less than or equal to
$\neq$	does not equal
α	is proportional to
$:$	is to; the ratio of
$\therefore$	therefore
$\cdots$	et cetera, i.e., $P = P_1 + P_2 + P_3 + \cdots$
Σ	the sum of, i.e., $P = \Sigma\, P_i$, means to add all the individual P_i values
∞	infinity
$\%$	per cent
$\bar{x}$	average value of x
σ	standard deviation
$!$	factorial $4! = 1 \times 2 \times 3 \times 4$
$\lvert x \rvert$	absolute value of x, without regard to sign
Δx	increment in x. $\Delta x = x_{\text{final}} - x_{\text{initial}}$
e	base of natural logarithms $= 2.718 \cdots$
$\ln x$	natural logarithm of x
$\log x$	base 10 logarithm of x
$\exp x$	e^x. $A \exp -\Delta E/RT = Ae^{-\Delta E/RT}$
$f(x)$	function of x
dx	differential of x
dy/dx	derivative of y with respect to x
$\int$	integral
$\int_a^b$	definite integral from a to b

APPENDIX 3

ANSWERS TO PROBLEMS AND EXERCISES

CHAPTER 1

PROBLEMS

1.1 (a) density = mass/volume; mass given, volume found by subtracting initial from final volume.

(b) A.W. = average mass of atom relative to atom of C-12. Divide A.W. by 12.000.

(c) 1 mole = 6.02×10^{23} molecules = formula weight in grams. Multiply by 6.02×10^{23} to get number of molecules, by 44.0 to get number of grams.

(d) $\Delta H_f = \Delta H$ when one mole of compound is formed from elements. Multiply given ΔH by formula weight of NiO, 74.7.

(e) Vapor pressure is independent of volume; pressure unchanged.

(f) m = number moles solute/number kg solvent; divide number of moles solute by 0.112.

1.2 (a) Same number of atoms of each element on both sides.

(b) $\Delta H = \Sigma \Delta H_f$ products $- \Sigma \Delta H_f$ reactants

(c) $P = nRT/V$; n = number grams $N_2/28.0$

(d) $\log \dfrac{P_2}{P_1} = \dfrac{\Delta H_{vap}(T_2 - T_1)}{(2.30)(1.99) T_2 T_1}$

(e) Number of moles $CaCl_2$ same before and after dilution.

(f) $\Delta S = (\Delta H - \Delta G)/T$

1.3 (a) grams Mg $\rightarrow$ moles Mg $\rightarrow$ atoms Mg

(b) grams H_2O $\rightarrow$ grams H; subtract from 1.60 g to obtain g of C. Find number of moles H, C by dividing number of grams by 1.01 and 12.0, respectively. Find simplest mole ratio of H to C, then simplest whole-number ratio.

(c) Calculate T from ideal gas law ($T = PV/nR$). Calculate $u = (3RT/M)^{1/2}$.

(d) Multiply atomic radius by 4 to find length of face diagonal, d. $l = d/\sqrt{2}$.

204

(e) Find $\Delta T_f = 5.50°C - T_f$. Calculate m from $\Delta T_f = 5.10$ m. Calculate M from equation: $m = \dfrac{\text{number g solute}/M}{\text{number kg solvent}}$

1.4 (a) ΔH = enthalpy change of reaction; ΔH_f = heat of formation of compound; Σ = sum of.

(b) P_{tot} = total pressure of gas mixture; $P_1, P_2, \cdots$ = partial pressures of components.

(c) P = pressure; V = volume; T = temp. in $°K$; n = number of moles; R = 0.0821 liter atm/mole $°K$.

(d) ΔT_b = bp elevation; k_b = bp constant of solvent; m = molality.

(e) ΔG = free energy change; ΔH = enthalpy change; ΔS = entropy change; T = temp. in $°K$.

(f) X_0, X = conc. at time 0 and t respectively; k = 1st order rate constant.

(g) K_b = ionization constant of weak base; K_a = ionization constant of conjugate weak acid.

1.5 (a) 1 and 2 wrong; combustion of propane is exothermic.

(b) All answers possible, depending upon amount of N_2.

(c) 3 wrong; can't reduce volume by dilution.

(d) 2 wrong; not a simplest formula.

(e) 2 very unlikely. (f) 1 wrong; molecular weight never less than 1.

1.6 (a) A.W. = 10.02(0.1883) + 11.01(0.8117) (b) atom ratio = $\dfrac{73.4/58.9}{26.6/16.0}$

(d) $V = \dfrac{(12.0)(0.0821)(300)}{(28.0)(750/760)}$ (e) $T = -3.10°C$

1.7 (a) Density water needed.

(b) Need abundance of one isotope (otherwise have more unknowns than equations).

(c) Don't need atomic weight of H.   (d) Pressure need not be constant.

(e) Volume of flask not necessary.   (f) Need ΔH_f (or ΔS) as well.

1.8 (a) How many grams of CO_2 are formed by the combustion of 1.60 mole CH_4?

(b) Calculate ΔH for the reaction: $2H_2(g) + O_2(g) \rightarrow 2H_2O(1)$

(c) A sample of a gas occupies 60.0 cc at 719 mm Hg and $21°C$. What volume will it occupy at 1.00 atm and $-22°C$?

(d) A solution of 1.60 g of a certain nonelectrolyte in 12.4 g of water freezes at $-1.34°C$. What is the molecular weight of the solute?

(e) The solubility of the compound MX_2 is 1.3×10^{-3} mole/liter. Calculate K_{sp}.

(f) K_a for the weak acid HA is 1.8×10^{-5}. Calculate K_b for A^-.

CHAPTER 2

PROBLEMS

2.1 (a) $40.6 \text{ g} \times \dfrac{1 \text{ kg}}{10^3 \text{ g}} = 4.06 \times 10^{-2} \text{kg}$ (b) $40.6 \text{ g} \times \dfrac{10^3 \text{ mg}}{1 \text{ g}} = 4.06 \times 10^4 \text{ mg}$

(c) $40.6 \text{ g} \times \dfrac{1 \text{ lb}}{454 \text{ g}} = 8.94 \times 10^{-2} \text{ lb}$ (d) $8.94 \times 10^{-2} \text{ lb} \times \dfrac{16 \text{ oz}}{1 \text{ lb}} = 1.43 \text{ oz}$

2.2   (a) $125 \text{ ml} \times \dfrac{1 \text{ liter}}{10^3 \text{ ml}} = 1.25 \times 10^{-1} \text{ liter}$

(b) $125 \text{ cm}^3 \times \left(\dfrac{1 \text{ m}}{10^2 \text{ cm}}\right)^3 = 1.25 \times 10^{-4} \text{ m}^3$

(c) $125 \text{ ml} \times \dfrac{1.057 \text{ qt}}{10^3 \text{ ml}} = 1.32 \times 10^{-1} \text{ qt}$

2.3   (a) $30.10 \text{ in Hg} \times \dfrac{25.40 \text{ mm Hg}}{1 \text{ in Hg}} = 764.5 \text{ mm Hg}$

(b) $764.5 \text{ mm Hg} \times \dfrac{1 \text{ atm}}{760 \text{ mm Hg}} = 1.006 \text{ atm}$

(c) $1.006 \text{ atm} \times \dfrac{1.013 \text{ bar}}{1 \text{ atm}} = 1.019 \text{ bar}$

2.4   (a) $3.12 \dfrac{\text{g}}{\text{cm}^3} \times \dfrac{1 \text{ kg}}{10^3 \text{ g}} \times \dfrac{10^6 \text{ cm}^3}{1 \text{ m}^3} = 3.12 \times 10^3 \dfrac{\text{kg}}{\text{m}^3}$

(b) $3.12 \dfrac{\text{g}}{\text{cm}^3} \times \dfrac{1 \text{ lb}}{454 \text{ g}} \times \dfrac{2.83 \times 10^4 \text{ cm}^3}{1 \text{ ft}^3} = 1.94 \times 10^2 \text{ lb/ft}^3$

2.5 $4.82 \times 10^4 \dfrac{\text{cm}}{\text{sec}} \times \dfrac{1 \text{ in}}{2.54 \text{ cm}} \times \dfrac{1 \text{ mile}}{6.34 \times 10^4 \text{ in}} \times \dfrac{3.60 \times 10^3 \text{ sec}}{1 \text{ hr}} = 1.08 \times 10^3 \dfrac{\text{mile}}{\text{hr}}$

2.6 $3.25 \text{ g} \times \dfrac{1 \text{ mole}}{16.0 \text{ g}} = 0.203 \text{ mole};$

$0.203 \text{ mole} \times \dfrac{6.02 \times 10^{23} \text{ molecules}}{1 \text{ mole}} = 1.22 \times 10^{23} \text{ molecules}$

2.7   (a) $1.60 \text{ moles} \times \dfrac{111 \text{ g}}{1 \text{ mole}} = 178 \text{ g}$

(b) $2.01 \times 10^{23} \text{ molecules} \times \dfrac{18.0 \text{ g}}{6.02 \times 10^{23} \text{ molecules}} = 6.01 \text{ g}$

2.8   (a) $1.51 \text{ moles NH}_3 \times \dfrac{5 \text{ moles O}_2}{4 \text{ moles NH}_3} = 1.89 \text{ moles O}_2$

(b) $0.282 \text{ moles NH}_3 \times \dfrac{6 \text{ moles H}_2\text{O}}{4 \text{ moles NH}_3} \times \dfrac{18.0 \text{ g H}_2\text{O}}{1 \text{ mole H}_2\text{O}} = 7.61 \text{ g H}_2\text{O}$

(c) $6.40 \text{ g NO} \times \dfrac{1 \text{ mole NO}}{30.0 \text{ g NO}} \times \dfrac{1 \text{ mole NH}_3}{1 \text{ mole NO}} = 0.213 \text{ mole NH}_3$

(d) $9.80 \text{ g O}_2 \times \dfrac{4(30.0)\text{g NO}}{5(32.0)\text{g O}_2} = 7.35 \text{ g NO}$

(e) From the equation: $68.0 \text{ g NH}_3 \triangleq 108 \text{ g H}_2\text{O}$

so: $68.0 \text{ lb NH}_3 \triangleq 108 \text{ lb H}_2\text{O}$

$1.00 \text{ lb H}_2\text{O} \times \dfrac{68.0 \text{ lb NH}_3}{108 \text{ lb H}_2\text{O}} \times \dfrac{0.454 \text{ kg}}{1 \text{ lb}} = 0.286 \text{ kg NH}_3$

2.9   (a) $4.08 - 2x$ (b) $3x$ (c) $28.0x$

2.10   (a) $1.00 \text{ g} \times \dfrac{213 \text{ kcal}}{16.0 \text{ g}} = 13.3 \text{ kcal}$

(b) 1.00 mole $O_2 \times \dfrac{213 \text{ kcal}}{2 \text{ moles } O_2} = 106 \text{ kcal}$

(c) 1.00 lb $CO_2 \times \dfrac{454 \text{ g}}{1 \text{ lb}} \times \dfrac{213 \text{ kcal}}{44.0 \text{ g } CO_2} = 2.20 \times 10^3 \text{ kcal}$

(d) O_2 is in excess (1 mole $O_2 \doteq 8.00$ g CH_4)

8.00 g $CH_4 \times \dfrac{213 \text{ kcal}}{16.0 \text{ g } CH_4} = 106 \text{ kcal}$

2.11 (a) $0.0821 \dfrac{\text{liter atm}}{\text{mole } ^\circ K} \times \dfrac{24.2 \text{ cal}}{1 \text{ liter atm}} = 1.99 \dfrac{\text{cal}}{\text{mole } ^\circ K}$

(b) $0.0821 \dfrac{\text{liter atm}}{\text{mole } ^\circ K} \times \dfrac{10^3 \text{ ml}}{1 \text{ liter}} \times \dfrac{760 \text{ mm Hg}}{1 \text{ atm}} = 6.24 \times 10^4 \dfrac{\text{ml mm Hg}}{\text{mole } ^\circ K}$

2.12 $1.99 \times 10^{-11} \dfrac{\text{ergs}}{\text{molecule}} \times \dfrac{1 \text{ kcal}}{4.18 \times 10^{10} \text{ ergs}} \times \dfrac{6.02 \times 10^{23} \text{ molecules}}{1 \text{ mole}}$

$= 287 \text{ kcal/mole}$

2.13 $12,100 \text{ coulombs} \times \dfrac{1 \text{ mole } e^-}{96,500 \text{ coul.}} \times \dfrac{1 \text{ mole } PbO_2}{2 \text{ moles } e^-} = 6.27 \times 10^{-2} \text{ mole } PbO_2$

$6.27 \times 10^{-2} \text{ moles } PbO_2 \times \dfrac{1 \text{ mole } PbSO_4}{1 \text{ mole } PbO_2} \times \dfrac{303 \text{ g } PbSO_4}{1 \text{ mole } PbSO_4} = 19.0 \text{ g } PbSO_4$

2.14 $\dfrac{112 \text{ mg } C_2H_5OH}{1 \text{ liter}} \times \dfrac{32.0 \text{ mg } O_2}{46.0 \text{ mg } C_2H_5OH} = 77.9 \dfrac{\text{mg } O_2}{\text{liter}}$

2.15 $1.00 \text{ g } D \times \dfrac{0.0256 \text{ g mass}}{4.02 \text{ g } D} \times \dfrac{2.15 \times 10^{10} \text{ kcal}}{1 \text{ g mass}} = 1.37 \times 10^8 \text{ kcal}$

CHAPTER 3

PROBLEMS

3.1 (a) 25.7; 0.257 (b) 3.89 lb
3.2 1st brand: $23/lb H_2O_2; 2nd brand: $19/lb H_2O_2
3.3 4.4×10^2 g
3.4 (a) 16.4 liters   (b) $273^\circ C$
3.5 54.9 g; 395 g
3.6 (a) 11.4 moles   (b) 16.2 moles
3.7 9.99; 0.0999 3.8 $0.800, 0.200$; 9.60
3.9 $75.5, 23.2, 1.2$ 3.10 55.86
3.11 $71\% Cu-63$; $29\% Cu-65$
3.12 (a) $36.1\% Ca, 63.9\% Cl$ (b) $3.08\% H, 31.6\% P, 65.3\% O$
    (c) $23.8\% Co, 33.9\% N, 3.66\% H, 38.7\% O$
3.13 (a) $NaSO_4$ (b) PH_4I
3.14 $1.3 \times 10^{-3}, 1.3 \times 10^{-3}, 0.099$
3.15 (a) 0.52 mole NO_2 consumed, 24% (b) $0.11 O_2, 0.22 NO, 0.68 NO_2$
3.16 40.6 3.17 2.4×10^2 g
3.18 (a) 5×10^{-5} (b) 0.5 (c) 5×10^2
3.19 (a) 7.3×10^{-7} (b) 6.8×10^4

CHAPTER 4

Section 4.1

1. (a) 1×10^3 (b) 1×10^9 (c) 1×10^{-6} (d) 1.622×10^4 (e) 2.126×10^2
   (f) 1.89×10^{-1} (g) 6.18×10^0 (h) 7.846×10^{-8}
2. (a) 3×10^3 (b) 10,000 (c) 2×10^{-4} (d) 4×10^8 (e) 1.5×10^{-2} (f) same

Section 4.2

1. 9.30×10^{12} 2. 3.9×10^2 3. 1.05×10^{-2} 4. 3.90×10^{16}
5. 1.26×10^3 6. 5.18×10^{-3} 7. 2.0×10^{-7} 8. 3.38×10^{15}
9. 1.48×10^{-18}

Section 4.3

1. 4.67×10^{-6} 2. 1.2×10^{14} 3. 3.6×10^{-41} 4. 3.0×10^3
5. 9.2×10^2 6. 2.5×10^{-1} 7. 4.02×10^{-4} 8. 2.78×10^3
9. 2.0×10^6 10. 3.1×10^1

Section 4.4

1. 4.71×10^4 2. 5.47×10^{-2} 3. 6.20×10^4 4. 2.45×10^{-4}
5. 8.05×10^5 6. 6.37×10^{-10} 7. 9.75×10^4 8. 6.02×10^{23}

Section 4.5

1. $3 \times 3 \times 3 \times 3 \times 3$ 2. 1   3. $1/3^2$ 4. $\sqrt[5]{6.4}$ 5. 3^{-2} 6. 8^3
7. 7^{-6} 8. $5^{3/2}$

PROBLEMS

4.1 6.65×10^{-24} g, 9.3×10^{-9} cm, 1.36×10^5 cm/sec
4.2 3.0×10^{-2} g/liter; 0.030 g/liter
4.3 9.43×10^4 cal   4.4 2.99×10^{-23} g
4.5 (a) 1.60×10^3 yr   (b) 5.84×10^5 days   (c) 8.41×10^8 min
4.6 (a) 4.0×10^{-8} g   (b) 5.0×10^6 liters
4.7 (a) 1.5×10^{-4} (b) 0.95
4.8 1.8×10^{-4}; 1.8×10^{-4}
     5.7×10^{-5}; 5.7×10^{-4}
     1.8×10^{-5}; 1.8×10^{-3}
4.9 5.3×10^{-9} cm; 2.1×10^{-8} cm; 4.8×10^{-8} cm

4.10 8.6×10^{-13} 4.11 5.57×10^{-8} cm
4.12 $3.2 \times 10^{-7}, 2.2 \times 10^{-3}, 2.1 \times 10^{-10}$
4.13 (a) 2×10^{-4} (b) 2×10^{-3}
4.14 2.18×10^4 cc   4.15 2.44×10^2 torr
4.16 $mole^{-2}$ $liter^2$ min^{-1}

CHAPTER 5

Section 5.1

1. 0.9117   2. 0.0128   3. 0.2180   4. 0.7665   5. 1.7522
6. 9.6212   7. $0.6918 - 3 = -2.3082$ 8. $0.5877 - 12 = -11.4123$

Section 5.2

1. (a) 7.640   (b) 8.644   (c) 55.40   (d) 41.35   (e) 4.189×10^7
2. (a) $0.3424 - 1$ (b) $0.5977 - 3$ (c) $0.3805 - 3$ (d) $0.6000 - 13$
3. (a) 0.2200   (b) 3.960×10^{-3} (c) 2.402×10^{-3} (d) 3.981×10^{-13}

Section 5.3

1. 2.0149 2. 2.1961 3. -2.5852   4. 8.8743   5. 0.1974
6. 19.2606 7. 1.392×10^{16} 8. 7.208×10^6 9. 14

Section 5.4

1. 1.796   2. 7.612   3. -9.693   4. 0.4343   5. 2.688
6. 9.33×10^9 7. 0.3679
8. $e = 2.000 + 0.500 + 0.167 + 0.042 + 0.008 + 0.001 = 2.718$;
    $\ln 2 = 2[\frac{1}{3} + \frac{1}{3}(\frac{1}{3})^3 + \frac{1}{5}(\frac{1}{3})^5] = 2(0.333 + 0.012 + 0.001) = 0.692$

PROBLEMS

5.1   (a) 6.00   (b) 1.583   (c) 8.52   (d) -0.78
5.2   (a) 1×10^{-4} (b) 2.5×10^{-13} (c) 7.2×10^{-4} (d) 10
5.3   2.872   5.4 1×10^{-15}
5.5   (a) 9.30   (b) 2×10^{-6}
       (c) log (conc H^+)(conc OH^-) = log (conc H^+) + log (conc OH^-)
$$= -pH - pOH = -14.0;$$

$$pH + pOH = 14.0$$

5.6   (a) 2.3×10^3 cal   (b) 5×10^{-3}
5.7   (a) 51 min   (b) 1.81 mole/liter
5.8   (a) 24 mm Hg   (b) $386°K$
5.9   11   5.10 12,800 cal
5.11 (a) -1.00 (b) 10^{-25}
5.12 4.77×10^{-8}
5.13 (a) 1   (b) 0.187   (c) 0.432
5.14 (a) 24 yrs   (b) 47 yrs

CHAPTER 6

Section 6.1

2. 1.06, 3.16, 6.87
3. 0.5%; note that the % error is approximately the same regardless of the section of the D scale you are using.

Section 6.2

1. 8.91   2. 24.8   3. 1.22   4. 0.217   5. 6.81   6. 0.611   7. 6.66

Section 6.3

2. 4.00×10^4 3. 9.49×10^8 4. 2.33×10^4 5. 4.55×10^4

Section 6.4

1. 2.86   2. 3.75×10^{-5} 3. 1.26   4. 9.33   5. 1.42×10^1
6. 6.21×10^3 7. 5.46×10^{-3}

Section 6.5

1. 2.25×10^2 2. 2.74×10^6 3. 2.11×10^{-10} 4. 2.35   5. 6.45
6. 1.25   7. 2.39×10^1 8. 8.40×10^{-2}

Section 6.6

1. 0.788   2. 5.111   3. -3.234 4. 1.037   5. 2.90   6. 1.78×10^1
7. 3.98×10^{-6}

PROBLEMS

6.1 3.30×10^{-5} 6.6   4.46   6.11 1.50×10^{-1} 6.16 1.07×10^{-1}
6.2 7.18×10^{-6} 6.7 2.24×10^{1} 6.12 3.87×10^{7} 6.17 2.99×10^{4}
6.3 9.49×10^{14} 6.8 2.15×10^{-66} 6.13 3.09×10^{2} 6.18   6.05
6.4   5.382   6.9 5.05×10^{3} 6.14   2.51   6.19 1.6×10^{7}
6.5 1.94×10^{-2} 6.10 5.76×10^{3} 6.15 2.69×10^{-6} 6.20 8.62×10^{7}

CHAPTER 7

Section 7.1

1. 4   3. 4   5. 5   7. 3   9. 1
2. 3   4. 3   6. 3   8. 2   10. 2 or 3

Section 7.2

1. 9.49×10^{-3} 2. 1.2×10^{2} 3. 1.02   4. 1.01×10^{-5} 5. 1.2
6. 1.68   7. 2.4   8. 5.80×10^{1}

Section 7.3

1. 15.33 g   2. 5.77 ml   3. 33.7 cm   4. 15.25 m   5. 1.06×10^{2} g
6. 3.34×10^{-1} ml   7. 75.2 cm^{2} 8. 0.52 lb/in^{2}

Section 7.4

1. 4.315465   2. 4.31546   3. 4.3155   4. 4.316   5. 4.32   6. 4.3
7. 4

Section 7.5

1. 0.2046   2. 0.72   3. 3.338   4. -3.31   5. 1×10^{2} 6. 1.53   7. 1.6×10^{3}
8. 1.25×10^{-2}

PROBLEMS

7.1   1.56 g/cc   7.2   2.2 g/cc
7.3   (a) 0.336 g   (b) 0.159 ml

7.4 617 g   7.5 9.995 ml
7.6 (a) 1.269 g   (b) 35.3% C, 2.5% H, 62.21% I
7.7 (a) 7.334 g   (b) 23 g
7.8 0.230   7.9 222 g/mole
7.10 (a) 10^{-4} (b) 8×10^{-5} (c) 7.6×10^{-5} (d) 7.62×10^{-5}
7.11 5542 cal   7.12 A.W. $= 2.00x + 34.97 = 35.45$

CHAPTER 8

Section 8.1

1. (a) 1.5×10^{-4} (b) 11/13   (c) 1.33   (d) 12/7   (e) 10.6   (f) 2.1
    (g) 4.4×10^{-4} (h) 0.71

2. (a) $P_1 V_1 / P_2$ (b) PV/nR (c) $(2E/m)^{1/2}$ (d) $\dfrac{1000}{18.0}\left(\dfrac{x}{1-x}\right)$

Section 8.2

1. 68% Al

Section 8.3

1. $-x$; $1.00 - x$; $\dfrac{x^2}{1.00 - x} = 7.0 \times 10^{-4}$

    $+x$; x
    $+x$; x

2. $-x$; $0.50 - x$; $\dfrac{x(0.10 + x)}{0.50 - x} = 7.0 \times 10^{-4}$

    $+x$; $0.10 + x$
    $+x$; x

3. $-2x$; $1.00 - 2x$; $\dfrac{4x^3}{(1.00 - 2x)^2} = 5.0 \times 10^{-4}$

    $+2x$; $2x$
    $+x$; x

4. $-2x$; $1.00 - 2x$; $\dfrac{4x^2(1.00 + x)}{(1.00 - 2x)^2} = 5.0 \times 10^{-4}$

    $+2x$; $2x$
    $+x$; $1.00 + x$

5. $-x$; $1.00 - x$; $\dfrac{4x^2}{27(1.00 - x)^4} = 5.0 \times 10^2$

    $-3x$; $3.00 - 3x$
    $+2x$; $2x$

6. $-x/2$; $1.00 - x/2$; $\dfrac{x^2}{27(1.00 - x/2)^4} = 5.0 \times 10^2$

$-3x/2$; $3.00 - 3x/2$

$+x$; x

Section 8.4

1. (a) $\pm 3.9 \times 10^{-5}$ (b) 0.59, 3.4   (c) 0.42, -0.72 (d) 0.36, 0.68
   (e) 1.5, -6.5 (f) 0.14, -0.15
2. (a) 4.24×10^{-3} (b) 3.44×10^{-2}

Section 8.5

1. (a) 1.0×10^{-3} (b) 1.0×10^{-2} (c) 6.3×10^{-3}
2. (a) 1.0×10^{-2} (b) 4.0×10^{-3} (2 appr.)   (c) 2.6×10^{-2}
3. (a) 2.0×10^{-2} (b) 0.98

Section 8.6

1. (a) $x = -1/33, y = -59/33$ (b) $x = 0.098, y = 0.102$
2. $x = 70/17, y = 7/34, z = 97/34$

PROBLEMS

8.1 $°F = 1.8°K - 459$ 8.2 $333°K$

8.3   (a) 5.5×10^{-27} g   (b) 6.0×10^{-36} cc

8.4   (a) 1.29   (b) 0.771   (c) 0.745

8.5 $d = PM/RT = 3.98$ g/liter   8.6   24.2 atm vs. 24.6 atm

8.7   (a) $m = \dfrac{c}{d - cM_2/1000}$ (b) 0.1014

8.8   34.5% LiCl

8.9   (a) $2 \times 10^{-13} = \dfrac{[Cu^{2+}] \times [NH_3]^4}{[Cu(NH_3)_4^{2+}]}$ (b) $1.0 \times 10^{-9} = \dfrac{[H_2]^2 \times [O_2]}{[H_2O]^2}$

     (c) $2.0 \times 10^4 = \dfrac{[N_2] \times [H_2]^3}{[NH_3]^2}$

8.10   (a) $2 \times 10^{-13} = 256x^5/(1.00 - x)$ (b) $1.0 \times 10^{-9} = 4x^3/(1.00 - 2x)^2$
    (c) $2.0 \times 10^4 = 27x^4/(1.00 - 2x)^2$

8.11   0.85, 0.85, 0.15

8.12   (a) 2.1×10^{-2} (b) 6.5×10^{-3} (2 appr.)   (c) 1.9×10^{-3} (2 appr.)

8.13   0.018; $(1.5 - 2x \approx 1.5 - x \approx 1.5)$

8.14 $[NH_4^+] = 1.0$; $[NH_3] = [H^+] = 2.4 \times 10^{-5}$

CHAPTER 9

Section 9.1

1. (a) 7/2   (b) 7   (c) 49/2   (d) −35/2   (e) 7/4
2. (a) 1/2   (b) 1   (c) 7/2   (d) −5/2   (e) 1/4
3. (a) $a(x_2 + x_1)$ (b) $a(x_2 - x_1)$ (c) a
4. (a) $y = ax$; 20   (b) $y = ax^2$; 32   (c) $y = ax$; 11/3   (d) $y = ax^{1/2}$; 4

Section 9.2

1. (a) 6.0, 2.0   (b) 300, 30, 0.30
2. (a) Inverse.   (b) Direct.   (c) Inverse. (d) Neither.
3. (a) 5   (b) 1/5   (c) −4
4. $(x_1/x_2)^2$

Section 9.3

1. (a) direct + linear   (b) linear   (c) linear
2. (a) $a = 1.2, b = 0$ (b) $a = 1.2, b = 1.0$ (c) $a = 2.5, b = -2.0$
3. (a) 3   (b) −18.0   (c) −4/3

Section 9.4

1. (a) $A = 1.00, B = 0$ (b) $A = 2.00, B = 3.00$
2. (a) 0.464   (b) 215
3. 1.00, 10.0, 1.31
4. (a) $\log y_2/y_1 = a(x_1 - x_2)$ (b) $y_2 - y_1 = a \log x_1/x_2$

Section 9.5

1. (a) $y = axz$ (b) $y = ax/z$ (c) $y = ax^2 z^{1/2}$ (d) $y = a/xz^2$
2. (a) Direct prop. to u, direct prop. to $v^{1/2}$, inversely prop. to z.

   (b) $y_2/y_1 = \dfrac{u_2 z_1}{u_1 z_2} \left(\dfrac{v_2}{v_1}\right)^{1/2}$ (c) $y_2/y_1 = u_2 z_1/u_1 z_2$

3. m n
   (a)   1   0
   (b)   1   1
   (c)   −1   1
   (d)   −1   0

Section 9.6

1. (a) 24   (b) $a = 1, b = 2$ (c) $a = 1, b = 2, c = -1$
2. 4.2

PROBLEMS

9.1   (a) 36.5 liters   (b) 0.536 mole

9.2   (a) $u = a(T/M)^{1/2}$ (b) $u_2 = u_1 \left(\dfrac{T_2 M_1}{T_1 M_2} \right)^{1/2}$

9.3 $a = nR/P$; (a) 0.0821 (b) 4.10

9.4 $T = t + 273$; $V = at + 273a$

9.5   0, 40, 80, 120, 200 mm Hg

9.6   (a) $fpl = ag_2/g_1 M_2$ (b) $2.00°C, 0.667°C$ (c) 278

9.7   (a) 1/2   (b) 48

9.8 $a = 1.34 \times 10^{-5}$; $b = 0.073554$

9.9   (a) -4.53×10^{-2} kcal/°K   (b) -24.4 kcal   (c) 1040°K

9.10 $\dfrac{[NO_2]^2}{[NO]^2 \times [O_2]} = 10.0, 10.0, 10.0, 10.0$

9.11 $n = 2, K = 10.0$ 9.12 $m = 1, n = 0, p = 2$

9.13   (a) mole^{-1} liter sec^{-1} (b) multiply by 60

9.14   (a) 0.26 atm   (b) 335°K   (c) 7370 cal

9.15 $A = 2200, B = 8.79$

9.16   (a) directly prop. to $[HI]^2$; inversely prop. to $[I_2]$
        (b) directly prop. to $[H_2]^{1/2}$ and $[I_2]^{1/2}$
        (c) inversely prop. to $[Cl^-]$
        (d) directly prop. to $[Ag(NH_3)_2{}^+]$; inversely prop. to $[NH_3]^2$
        (e) directly prop. to $[Ag^+]$; directly prop. to $[NH_3]^2$
        (f) directly prop. to $[H^+]$ and $[F^-]$

CHAPTER 10

Section 10.1

1. +2; +22 kcal/mole   2. $D: x = 5, y = 8$; $E: x = 6, y = 9.5$
3. (a) 9 mm Hg   (b) 29°C

Section 10.2

1. (a) to (d) See graphs pp. 216–219.

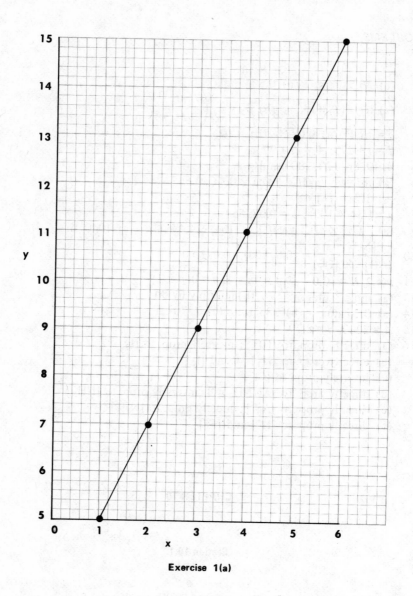

Exercise 1(a)

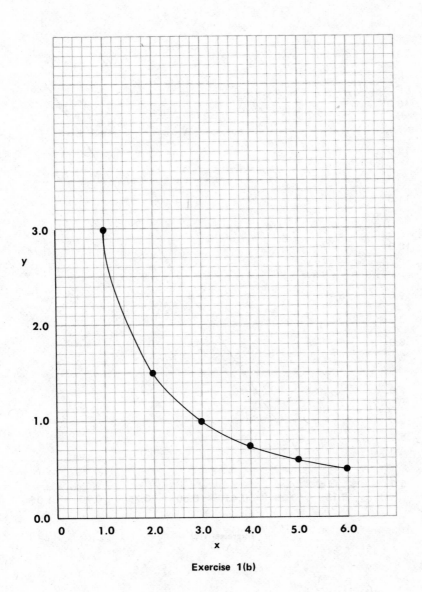

Exercise 1(b)

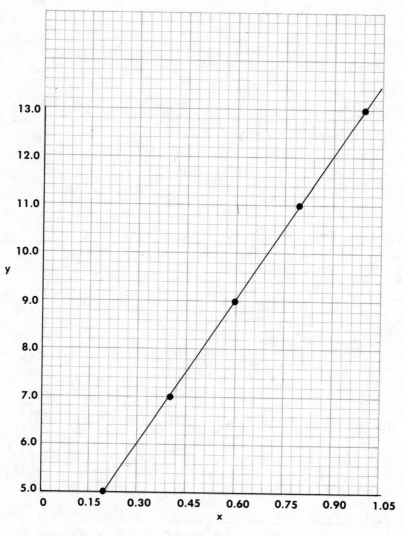

Exercise 1(c)

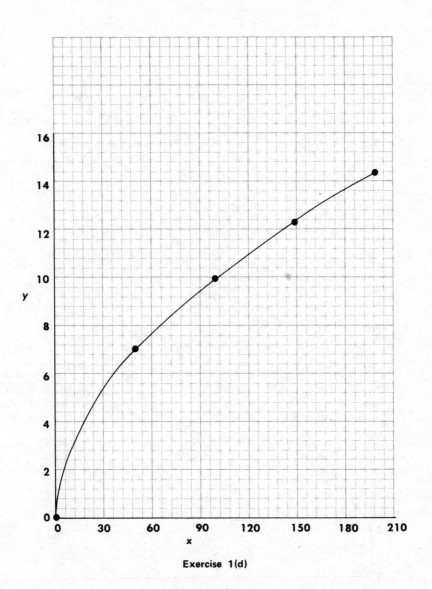

Exercise 1(d)

Section 10.3

1. (a) to (d) See graphs pp. 220–223.
2. See graph, p. 224.

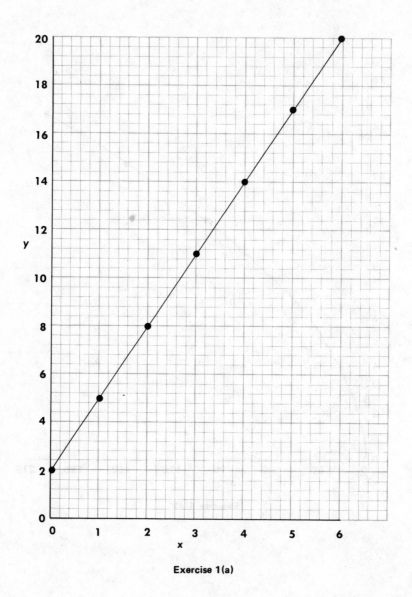

Exercise 1(a)

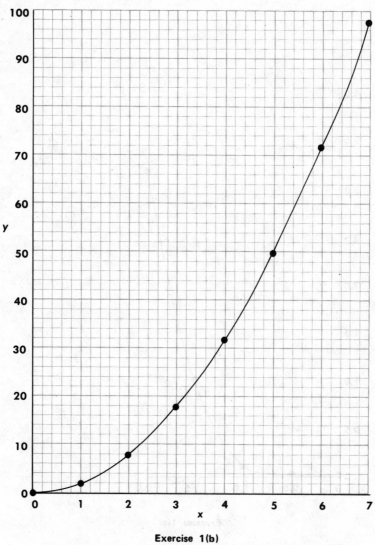

Exercise 1(b)

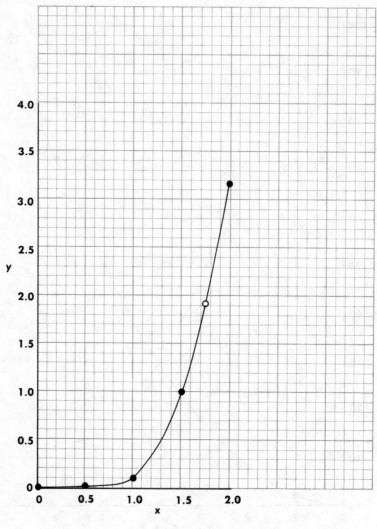

Exercise 1(c)

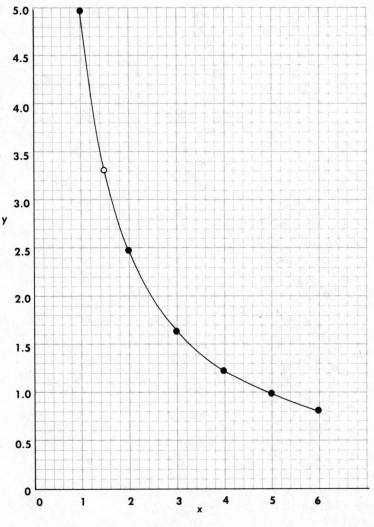

Exercise 1(d)

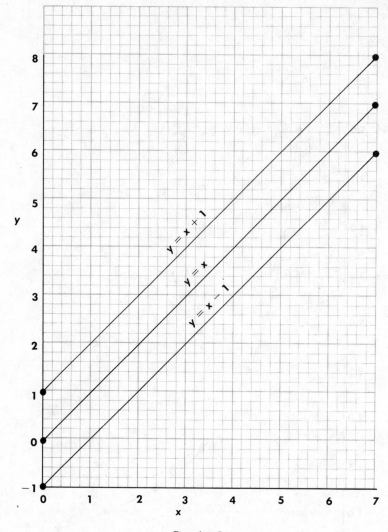

Exercise 2

Section 10.4

1. (a) $\log y$ vs. x (b) y vs. x^2 (c) y/x vs. x
2. (a) $a = 2.0, b = -6.0$ (c) $a = 0.51, b = 1.0$
   (b) $a = 3.0, b = -0.1$ (d) $a = -2.1, b = 2.9$
3. $A = 2.46, B = 0.50$

PROBLEMS

10.1 See graph, below.

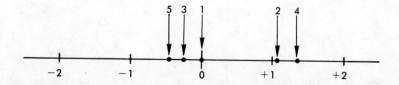

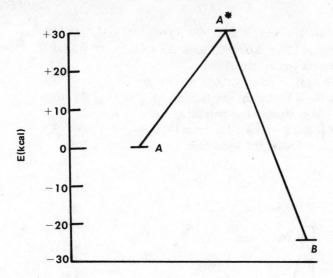

10.2   See graph, above.
10.3   P   5   7   9   13   18   24   32   42   55   72   93
       T   0   5   10   15   20   25   30   35   40   45   50
10.4   (See graph, below).   (a) −5.6 kcal   (b) −12.7 kcal   (c) 470°K

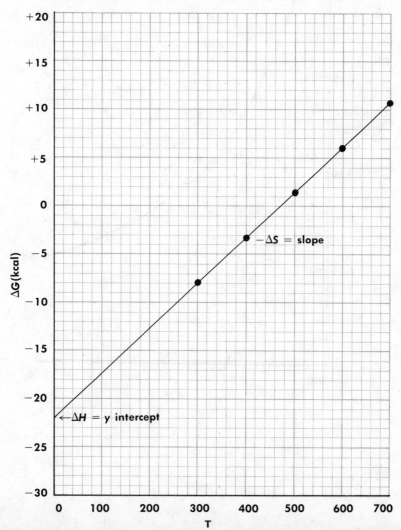

10.5   Straight line from (−33.4 kcal, 300°K) to (−1.7 kcal, 1000°K).
10.6   (See graph, p. 225.) $\Delta H = -22$ kcal, $\Delta S = -4.7 \times 10^{-2}$ kcal/°K
10.7   (a) hyperbola   (b) straight line
        (c) (conc Ag^+)$^2 \times$ (conc CrO_4^{2-}) = constant
10.8   (a) $P = 0.0082\ T$; straight line from (1.64, 200) to (3.28, 400)
        (b) $PV = 24.6$; straight line parallel to x axis
        (c) $PV = 0.082\ T$; straight line from (0, 0) to (98, 1200)
10.9   (See graph, below.) (a) hyperbola

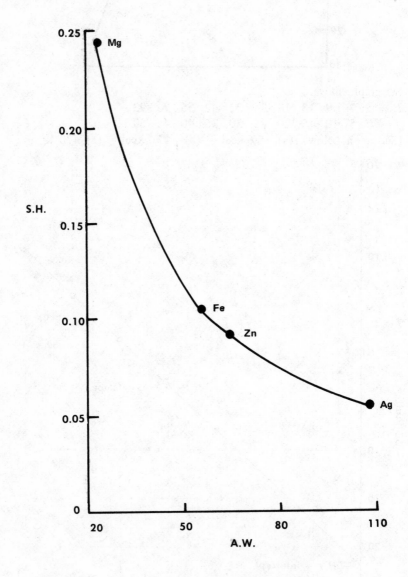

(b) (See graph, below.) plot S.H. vs. 1/A.W.   (c) 1/A.W. = 0.0250;
S.H. = 0.145

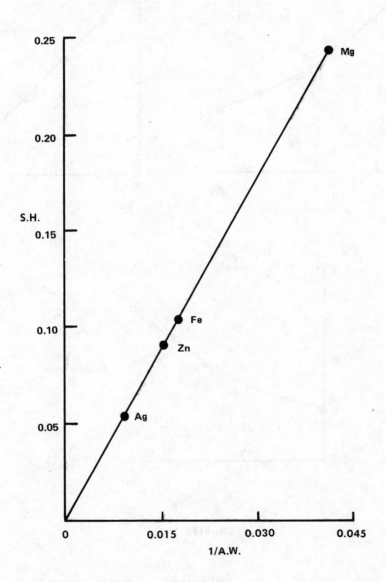

10.10 $a = -0.08, b = 1.0$

10.11   slope $\approx -1.67 \times 10^3$; $\Delta H \approx 7.64 \times 10^3$ cal

10.12   intercept = 2.42 = $\dfrac{-\log K_a}{2}$; $K_a = 1.5 \times 10^{-5}$

10.13 (See graph, p. 00.) 1st order. Slope ≈ -0.10; $k = 0.23$

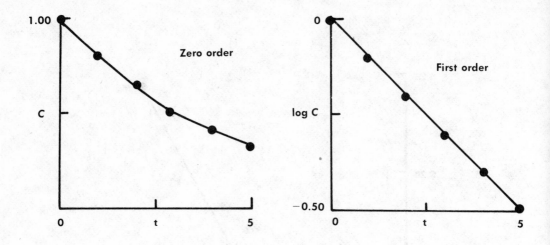

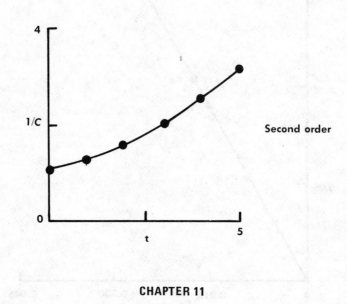

CHAPTER 11

Section 11.1

1. Observed T	Error	% Error	Deviation	% Deviation
328.1	+0.9	+0.3	+0.5	+0.2
327.6	+0.4	+0.1	0.0	0.0
327.3	+0.1	+0.03	-0.3	-0.09
327.4	+0.2	+0.06	-0.2	-0.06

Note that changing to °K changes only the % error and % deviation.

2. Mean = 2.3102

Trial	1	2	3	4	5	6	7	8	9
Error	+.0105	+.0102	+.0104	+.0103	+.0105	+.0104	+.0106	+.0119	+.0099
Deviation	.0000	-.0003	-.0001	-.0002	.0000	-.0001	+.0001	+.0014	-.0006

Section 11.2

1. (a) C too small; GEW too large   (b) C too large; GEW too small
   (c) C too large; GEW too small
2. (a) 28.0 g   (b) 22.4   (c) -20%
3. (a) 1.4   (b) 5
4. (a) 2050Å; +4Å(1); +3Å(1); +2Å(2); +1Å(4); 0Å(5); -1Å(3); -2Å(2); -3Å(1); -5Å(1)
   (b) Plot numbers in parentheses above vs. magnitude of deviation in Å. Curve resembles those in Figure 11.1, but is less smooth.

Section 11.3

1. $a = 0.036$; $\sigma = 0.052$
2. (a) Mean = 112.30; $\sigma = 0.060$ (b) about 2/3
3. (a) $y = 1/2.5\,\sigma$
   (b)

y	0.05	0.13	0.24	0.35	0.40	0.35	0.24	0.13	0.05
x	-2.0	-1.5	-1.0	-0.5	0.0	+0.5	+1.0	+1.5	+2.0

   Plot resembles Figure 11.1(a).
   (c)

y	0.00	0.01	0.11	0.49	0.80	0.49	0.11	0.01	0.00
x	-2.0	-1.5	-1.0	-0.5	0.0	+0.5	+1.0	+1.5	+2.0

   Plot resembles Figure 11.1(b).

Section 11.4

1. $M = 112.30 \pm 0.060t/\sqrt{6}$; ±0.018, ±0.049, ±0.099
2. ±4.5, ±2.3, ±0.10
3. $\sigma = 0.42$; $M = 16.9 \pm 64(0.42)/\sqrt{2} = 16.9 \pm 19$
   If he wants to be 99% sure, he had better repeat the experiment.
4. (a) ±4.5 (b) ±1.7 (c) ±1.2 (d) ±0.95 (e) ±0.82

Section 11.5

1. (a) Trial mean = 152.3; $a = \pm1.8$
   $4.3 < 2.5(1.8) = 4.5$; retain
   $4.3 < 4.0(1.8) = 7.2$; retain
2. 47.0, 49.4, 49.6, 50.1, 50.5, 55.2
   Test 55.2: $Q = 4.7/8.2 = 0.57 > 0.56$; reject
   Test 47.0: $Q = 2.4/3.5 = 0.69 > 0.64$; reject
   Test 50.5: $Q = 0.4/1.1 = 0.36 < 0.76$; retain

Section 11.6

1. (a) 2.8×10^{-3} g   (b) 2.8×10^{-3} g   (c) ± 0.09

PROBLEMS

11.1 Too large in each case.
11.2 (a) 36.1
     (b) $-0.1, -0.3\%; +0.2, +0.6\%; -0.3, -0.8\%; +0.2, +0.6\%$
     (c) 0.20
     (d) 0.24
11.3 $\pm 0.03, \pm 0.08, \pm 0.17$
11.4 Mean = 20.10; $\sigma = 0.10$; $M = 20.10 \pm 0.06$
11.5 1.08, 1.10, 1.12, 1.21
     Test 1.21: $Q = 0.09/0.13 = 0.69 < 0.76$; retain
11.6 39, 56, 58, 60, 60, 61, 62, 62, 64, 69
     Test 39: $Q = 17/30 = 0.57 > 0.41$; reject
     Test 69: $Q = 5/13 = 0.38 < 0.44$; retain
11.7 (a) $E_S = 0.14$ kcal/$300°K = 4.7 \times 10^{-4}$ kcal/$°K$
     (b) $E_G = (0.088)^{1/2} = 0.3$ kcal
11.8 $M = 155 \pm 0.6$

INDEX